Seaway to the Future

❖❖ STUDIES IN AMERICAN ❖❖
THOUGHT AND CULTURE

Series Editor

Paul S. Boyer

Observing America:
The Commentary of British Visitors to the United States,
1890–1950
Robert P. Frankel

Picturing Indians:
Photographic Encounters and Tourist Fantasies
in H. H. Bennett's Wisconsin Dells
Steven D. Hoelscher

Cosmopolitanism and Solidarity:
Studies in Ethnoracial, Religious, and Professional Affiliation
in the United States
David A. Hollinger

Seaway to the Future:
American Social Visions and the Construction of the Panama Canal
Alexander Missal

Unsafe for Democracy:
World War I and the U.S. Justice Department's Covert Campaign
to Suppress Dissent
William H. Thomas Jr.

Seaway to the Future

*American Social Visions and
the Construction of the Panama Canal*

Alexander Missal

THE UNIVERSITY OF WISCONSIN PRESS

This book was published with support from
the Graham Foundation for Advanced Studies in the Fine Arts
and from the Anonymous Fund of the College of Letters and
Science at the University of Wisconsin–Madison.

The University of Wisconsin Press
1930 Monroe Street, 3rd Floor
Madison, Wisconsin 53711-2059

www.wisc.edu/wisconsinpress/

3 Henrietta Street
London WC2E 8LU, England

1   3   5   4   2

Library of Congress Cataloging-in-Publication Data
Missal, Alexander.
Seaway to the future : American social visions and
the construction of the Panama Canal / Alexander Missal.
p.   cm. — (Studies in American thought and culture)
Includes bibliographical references and index.
ISBN 978-0-299-22940-5 (cloth: alk. paper)
1. Panama Canal (Panama) — History.
2. Panama Canal (Panama) — In popular culture.
3. Canals — Social aspects — Panama.
4. Canals — Design and construction — Panama.
5. Panama Canal (Panama) — Social policy.
I. Title. II. Series.
F1569.C2M57   2008
972.87′5 — dc22   2008011969

For Leah Silverstein

Contents

Illustrations

Acknowledgments

"What did you do, come by way of the Panama Canal?"
Alvy Singer (alias Woody Allen)
in the movie *Annie Hall* wonders why
his girlfriend arrives late for a movie

Originally conceived as a shortcut, the Panama Canal does not always serve this purpose, as Woody Allen's character Alvy Singer and I can testify. After many years of discussion, research, and writing, I am pleased to be able to thank the people and institutions instrumental in the making of this book, which is the published version of a Ph.D. dissertation accepted by the Faculty of Arts at the University of Cologne, Germany. I am first of all indebted to my adviser, Norbert Finzsch, who taught me how to pursue the study of history with scholarly discipline and methodological curiosity reaching beyond the boundaries of our own field, for his constant encouragement during the evolution of this work. I am also grateful to James Gilbert, who guided the first steps of the project and was involved in many other stages, for his support as a mentor, and for many inspiring discussions about cultural and intellectual history.

I would like to thank the German Historical Institute (GHI) in Washington, D.C., and its staff, especially Christine von Oertzen, Dirk Schumann, and Christof Mauch, for sponsoring and assisting my research in Washington and San Francisco. I am also much obliged to my employer, the Deutsche Presse-Agentur (dpa), and especially to Michael Ludewig and Werner Scheib, for granting me two sabbaticals for this project. I would like to thank the staff of the Library of Congress in Washington, D.C., the National Archives and Records Administration (NARA) in College Park, Maryland, the Bancroft Library at the University of

California in Berkeley, the San Francisco History Center at the San Francisco Public Library, the Württembergische Landesbibliothek in Stuttgart, and many other libraries in Germany and the United States. In particular, I thank Josef Schwarz at NARA and David Kessler at Bancroft for their assistance. It has been a lifetime privilege to be able to spend a number of months in the Main Reading Room at the Library of Congress, encompassed with the aura of scholarship (or what I have been happy to construct as such).

I want to express my gratitude to numerous colleagues, among them Janet Davis, David E. Nye, and Jordan Goodman, who have engaged me in lively discussions in classrooms and cafés and at academic gatherings such as the Netherlands American Studies Association conference in Middelburg and the HistorikerInnentreffen der Deutschen Gesellschaft für Amerikastudien in Tutzing. I am also indebted to Carsten Peters, Charlotte Marr, James Gilbert, and Jens Jäger for their willingness to read the manuscript or parts of it and share their suggestions and critique with me, and to Hanjo Berressem, who agreed to evaluate the original thesis as a second reader. Matt Matsuda and Jürgen Martschukat have been role models for me as scholars, counsels, and good spirits. I also would like to thank my peer reviewers, the staff at the University of Wisconsin Press, and, last but not least, my editors Gwen Walker and Paul S. Boyer for their enthusiasm and support. I could not have asked for a better team to see this project through its final round.

I am forever grateful to my mother, Hannelore Missal, for investing all of her heart and mind into my education and for her outright support of the choices I made. Finally, I want to thank my friends and companions on both sides of the Atlantic whose hospitality I have enjoyed, who have guided me with unwavering interest and cheerful endurance, and who have given me their love, understanding, and encouragement (and sometimes all of the above): Annie, Kirk, Grace, James, and Lydia Mitchell, Jeanne Hardy and Bruce Baird, Andy Bopp, Amy Rubin and Stefan Knerrich, Stacey Philpot and Steve Buvel, Johann Reidemeister, Almut Pohl and Carsten Peters, Charlotte Marr, Annegret Wirths, and Brigitte Biehl.

Seaway to the Future

Approaching the Panama Canal

An Introduction

The Panama Canal, at least from a Western perspective, is a feature of world history. It symbolizes a century-old dream, the passage to India, connecting oceans and continents. At the time of its completion in 1914, it was an engineering feat of unique and mind-boggling dimensions, and almost a hundred years later—perhaps to the surprise of its makers—the original technology has largely remained in place and operation. Only in the fall of 2006 did a substantial expansion of the Canal get under way when the people of Panama approved of President Martín Torrijos's plans after much controversy.

The Panama Canal was built during a momentous era. The entry of Western societies into the modern age and the brutal colonization of other parts of the world was characterized by a vision of progress that two world wars with more than sixty million deaths would soon turn into a nightmare. The United States became involved in its conception, oversaw its construction, and controlled the waterway and the surrounding territory for decades. Panama was a subject of American foreign relations throughout the twentieth century. The public debates in the 1970s, the military intervention in 1989, and, finally, the return of the Canal to Panama on New Year's Eve of 1999 serve as the more recent examples of this continuity.

During a symbolic ceremony two weeks before the official transfer, former president Jimmy Carter pronounced a simple "it's yours" when he handed over the Canal Zone and the seaway itself to the people of Panama[1]—an unusual act of humility considering the direction American diplomacy has taken in the past years. In the light of a century of troubled relationships between Panama and the United States, his demeanor stands out as even more exceptional. During Carter's presidency, the new Canal treaties had been signed and ratified, after more than ten years of negotiations and a fierce political debate dividing the American public. The preceding Canal treaty dated back to 1903, when a lucky coincidence enabled President Theodore Roosevelt to witness the birth of the new nation of Panama and close a deal on the construction of a canal that turned the young Latin American country into a dependency. In 1906 Roosevelt took a trip to the Isthmus, the narrow strip of land between two oceans, and had his picture taken atop a ninety-five-ton steam shovel, a symbol of the determination and ingenuity of the American nation. His and Carter's visits could not have been more different.

This book is a study of Roosevelt's times, when American society, coping with the consequences of rapid social change, was engaged in a "search for order." Confronted with these challenges, policymakers and commentators viewed the Canal as an imperial project of building a new nation, a future America. Their interpretations, overlooked by historians in the past, form the basis of this analysis. The Panama Canal was a construction site, filled not only with mud and water but with meanings as well.

Toward a New Empire: Historiographical Perspectives

In 1993 the historians Amy Kaplan and Donald Pease edited a collection of essays entitled *Cultures of United States Imperialism*. In the introduction, Kaplan notes a missing link in American historiography, "the absence of culture from the history of U.S. imperialism; the absence of empire from the study of American culture." She continues: "The study of American culture has traditionally been cut off from the study of foreign relations. From across this divide, however, the fields of American studies and of diplomatic history curiously mirror one another in their respective blind spots to the cultures of U.S. imperialism."[2] There were exceptions to Kaplan's assessment, such as Robert Rydell's important study *All the World's a Fair: Visions of Empire at American International Expositions, 1890–1945*,[3] but generally speaking, her analysis and the essays included in the 1993 volume articulated (and inspired) an overdue interest in the cultural history of empire that has not subsided since. Even though the building of the Panama Canal was alluded to in *Cultures of United States Imperialism* in a provoking and encouraging essay by Bill Brown,[4] one of the largest (if not the largest) imperial projects in American history has yet to be fully included in this debate. I believe that a new look at the Canal and its interpretations offers valuable insights into the profound changes American society was faced with at the turn of the twentieth century.[5]

For many decades, the study of American foreign relations and the nation's rise to global power had been dominated by two different strands of scholarship. After World War II, the protagonists of the Realist School, exemplified by the diplomat and historian George F. Kennan and his study *American Diplomacy, 1900–1950*,[6] believed that throughout the early decades of the century, the diplomacy of American leaders had been too incoherent and dependent on public opinion, preventing a clearly defined Realpolitik designed by experts, which would have better served the national interest. In 1959 William Appleman Williams refuted this claim in his influential work *The Tragedy of American Diplomacy*.[7] He argued that the United States had indeed pursued a consistent foreign policy: its expansion abroad was based on political and economic interests, but unlike the European powers' annexation of foreign territories, American influence was mostly (though not exclusively) exerted informally. It is important to note that Williams viewed domestic and foreign policies as separate but inherently related fields: economic crises at home prompted businessmen and politicians to search for new markets in other countries and devise strategies of expansion. Williams's thesis was elaborated by his students and other representatives of the New Left, most distinctly in Walter LaFeber's 1963 study *The New Empire: An Interpretation of American Expansion, 1860–1898*.[8] LaFeber emphasized the strategic and mercantile goals of diplomacy. Similar "dichotomous political-economic models,"[9] sometimes reinforced by methodological concepts such as the dependency theory, remained dominant until well into the 1980s.[10]

Many aspects of this development are reflected in the historiography of the Panama Canal. It commences with a long list of monumental studies stressing the epic nature of the Canal building, culminating in David McCullough's seven-hundred-page volume *The Path between the Seas: The Creation of the Panama Canal, 1870–1914*.[11] McCullough's book is not only highly readable and entertaining but also provides a wealth of details that had never been researched before. On the other hand, it offers almost no references to the domestic context of the Canal project and also lacks a critical analysis of U.S. policy toward Panama. The latter was provided by none other than Walter LaFeber in his study *The Panama Canal: The Crisis in Historical Perspective*, a book written in response to the debates of the 1970s.[12] Recently, Matthew Parker tried to step into McCullough's footsteps with his study *Panama Fever: The Battle to Build the Canal*,[13] but beginning with LaFeber, almost all general works published on the waterway since the 1970s were syntheses focusing on the evolution of the political issues surrounding Panama and the Canal.[14] Consequently, the circumstances of its construction, especially Roosevelt's "taking" of the Canal Zone in 1903, were primarily discussed with respect to the strained relations between Panama and the United States in the twentieth century. Even LaFeber, who, in Williams's tradition, had described the domestic origins of early American expansionism in *The New Empire*, dropped this link in *The Panama Canal*. The seaway had become a (geo)political subject in the context of American foreign relations, and not much else.[15]

Traditionally, diplomatic history has been practiced mostly by men, and this may be one of the reasons why aspects such as gender or "culture" have not been considered worthy of study. They were, as Amy Kaplan suggests, associated with the domestic realm.[16] Among others, Akira Iriye and Emily Rosenberg began to broaden the scope of the history of foreign relations by adding culture to their analysis, which was often based on Williams's and LaFeber's approach.[17] The foundation for a more fundamental shift in the practice of the discipline—visible for the past fifteen to twenty years—was then laid by the *linguistic* or *cultural turn* in the humanities, resulting in an inadequately but effectively named "new cultural history."[18]

Informed by postmodern and poststructuralist theories, this approach assumes that history is shaped not only by the exertion of physical or structural power through individuals, groups, or institutions but by discourses as well. These are determined by texts, images, symbols, and—a clumsy term—"material practices" revolving around a certain subject, producing and reproducing its interpretations, shaping and reshaping the attitudes and actions of the people concerned with and affected by it. Categories such as gender, race, or nation are not stable but socially constructed, contradictory domains, discursive sites of power and resistance at the center and the periphery of societies. I clarify my own understanding and use of the term "discourse" in a subsequent section and focus here on its historiographical relevance. The new cultural history is an integrative, not an "add-on," approach, going beyond the exercise of simply attaching aspects of gender, cultural, or intellectual history to the analysis of diplomatic history and other fields. In this sense, Williams's informal U.S. empire can be described—never mind the postmodern jargon—as "a chain of nodes or points of textual/ image production, each involving a combination of technologies, practices, and forms; as an extended flow of information, visual images, arguments, and meanings going from North to South; as a process of accumulation of symbolic capital through multiple technologies of seeing, narrating, and displaying."[19]

This cultural turn has added a new complexity to the study of foreign relations, transcending the confines of the traditional imperialist history that mainly encompassed Western nations and their colonial politics. It has given rise to manifold developments in the practice of history: First, the attempt to fill the gap that Amy Kaplan has observed—to link foreign policy and its interpretations to the corresponding domestic processes, to connect "empire building" and "nation building." This endeavor is the main focus of this book. Second, numerous efforts have been and will be made to extend the field beyond national boundaries to postcolonial and transnational topics.[20] Third, a debate revolving around the terms "imperialism" and "empire" is in full swing—in fact, "empire" has become ubiquitous.[21] One aspect of this latter discussion, inspired by the transnational approach, focuses on the question whether the foreign policy of the United States—itself a former colony—was inherently different from the diplomacy of European "imperialist" powers, or whether a constant exchange and assimilation of ideas and practices took place.[22] Such an approach must address the question of American "exceptionalism," either with an affirmative or a revisionist purpose.

There is another aspect to this debate on empire, based on the important work by Michael Hardt and Antonio Negri.[23] Empire is the discursive concept these scholars use to explain the forces of globalization. On the one hand, it is not attached to a specific nation but viewed as a global expansionist impulse without center, exerted through a web of institutional and social forces, including, for instance, multinational corporations. On the other hand, the evolution of Hardt's and Negri's empire does in fact go back to the creation of the United States. In contrast to other powers that tried to expand abroad and yet contain their rule by sovereign administration, the American nation, the authors argue, is based on the idea of a limitless space, of "order always renewed and always re-created in expansion," growing out of the vision of the U.S. constitution and the settlement of the American West.[24] The fact that the United States remains the only superpower of the twenty-first century underscores the connection between American expansionism and Hardt's and Negri's perspective on the present global empire. While I agree with their general argument, I would like to contest one of their historical assessments: Whereas Hardt and Negri state that Theodore Roosevelt "exercised a completely traditional European-style imperialist ideology" and only President Woodrow Wilson continued the Jeffersonian legacy of expanding networks,[25] I show that the Panama Canal, as conceived by Roosevelt and other expansionists, already prefigured the postmodern concept of empire. I also employ the term "empire" because it was used by contemporary policymakers and authors.[26] For them, empire encompassed more than territorial expansion; it implied a new way of life in the twentieth century, permeating all levels of society. The Panama Canal was the prime example of this process: on the Isthmus, the modern American nation was under construction.

Despite the neglect of the Panama Canal in the new cultural histories of empire, valuable contributions to specific aspects of the construction period have been made during the past twenty years, notably on environmental[27] and labor history.[28] All of these works have shed light on the racist regime erected in the Canal Zone. Julie Greene's forthcoming study *The Canal Builders: Making America's Empire at the Panama Canal* will presumably extend this discussion and also address some of the issues on which I focus in this book. The cultural role of technology in the evolution of American imperialism in Panama and elsewhere has been addressed as well.[29] And yet the literature on the interpretations of the Canal building in the United States remains sparse, even though Alfred Charles Richard noted in his study *The Panama Canal in American National Consciousness, 1870–1990* that the engineering feat was one of the most popular subjects of its time: "No other topic of public concern in the period 1903–1915 prompted so great a volume of printed material as that generated by the events on the isthmus."[30] Richard and J. Michael

Hogan, in his book *The Panama Canal in American Politics: Domestic Advocacy and the Evolution of Policy*, both address the issue of explaining the interoceanic canal, from its conception to its construction, to the American public.[31] Apart from the policymakers directly involved in this process—the traditional protagonists of diplomatic history—they quote a different group of authors as their sources: journalists, travel writers, Canal officials, and other people involved with the project who wrote magazine articles and popular books on the building of the waterway, the majority of which were published in the years prior to the opening of the Canal in 1914. These authors, whom Hogan calls the "storytellers,"[32] were instrumental in forging the public discourse on the Panama Canal, and I generally refer to them as the "Panama authors."[33] Hogan acknowledges their major role in depicting the Canal building as an American success story, and the term "story" is indeed appropriate since what they wrote must be viewed as a highly constructed narrative rather than objective reporting or even historical "fact." I show that this discourse provided complex interpretations of the Canal, going far beyond symbolizing "the material and spiritual benefits of international adventurism" for which Hogan credits these authors along with politicians such as Theodore Roosevelt.[34] While Richard and Hogan examine the public attention during the construction era in the context of U.S. attitudes toward Panama and the Canal throughout the twentieth century, I am interested in the social and cultural background of the authors' interpretations at the time when American society was at the crossroads between the Gilded Age and the modern era. For the authors, the Panama Canal was a utopian experiment, a showcase of the future America. So far, the only study that has addressed this relationship is not the work of a cultural or diplomatic historian but an autobiographical sketch by two former teachers on the Isthmus, Herbert and Mary Knapp's *Red, White and Blue Paradise: The American Canal Zone in Panama*.[35] With regard to the Panama Canal, Amy Kaplan's assessment of "the absence of culture from the history of U.S. imperialism; the absence of empire from the study of American culture" still rings true. As an "expressly artificial addition to the nation,"[36] the Canal Zone was neither part of the United States nor a proper colony, neither foreign nor domestic. This may explain why it has so far eluded the practitioners of a new cultural and diplomatic history. But it is precisely this terrain of "ambiguous spaces"[37] that makes the Panama Canal and its interpretations relevant to historical inquiry.

The Search for Order: The Canal and Its Interpreters

In all industrializing nations, the late nineteenth and early twentieth centuries were times of profound change. The United States, a rising economic superpower, was no exception. Along the lines of gender, race, and class, the participants in this process were raising their voices, articulating new identities, and demanding a share of the wealth that was accumulating in the hands of the upper class. Women fought for their right to vote and assumed new roles in a society increasingly shaped by a culture of consumption. African Americans migrated in large numbers from the South to the urban centers in the North, attracted by jobs at large factories. Beginning in the 1890s, millions of immigrants from Southern and Eastern Europe, many of them single males of Jewish or Roman Catholic faith, moved to the big cities, challenging the traditional pecking order. The metropolises became sites of conflict not only between the rich and the poor but also among the poor themselves, resulting in social unrest and racial violence. Depressions hit the economy. Within a period of fifteen years, as Gail Bederman notes, almost thirty-seven thousand strikes broke out, involving as much as quarter of the entire workforce.[38]

While the violent outbursts of social transformation had largely subsided by the turn of the century, a general sense of uneasiness prevailed. At the dawn of the modern era, advances in technology and communication affected all levels of society, public and private life. Confronted with a new environment, people needed to comprehend and cope with the rapid changes around them. Fear of the unfamiliar, of chaos, of being left behind was a constant undercurrent of these times. "We are homeless in a jungle of machines and untamed powers that haunt and lure the imagination. Of course, our culture is confused, our thinking spasmodic, and our emotion out of kilter," the American journalist Walter Lippmann wrote. "No mariner ever enters upon a more uncharted sea than does the average human being born into the twentieth century. Our ancestors thought they knew their way from birth through all eternity; we are puzzled about the day after to-morrow."[39] Despite this anxiety and confusion, many middle-class Americans expressed an optimistic outlook, however superficial it may have been. The unprecedented social and technological progress also held the promise of greater wealth and opportunities at home and abroad, the advance of civilization, even of a utopian future. For a nation built on the dream of a new society—from which others, such as African and Native Americans, were brutally excluded—this was a well-known concept.

To overcome fear and escape chaos, reformers proposed a new matrix for society, redefining the relationships between the individual and the state, men and women, between workers and their masters. Historians usually speak of the "Progressive Era," "Progressivism," and the "Progressive Movement." Even though Peter Filene and Daniel Rodgers have convincingly argued against the coherence of this "movement," the term "Progressivism" survives, if only as a common label. "Like an unpleasant party guest," James Connolly notes, "Progressivism refuses to go away."[40] People as diverse as the politician and amateur hunter Theodore Roosevelt, the intellectual Lippmann, and social reformer Jane Addams were all counted as Progressives. I prefer Robert Wiebe's title *The Search for Order* to describe the sum of these efforts. While its protagonists had divergent backgrounds and motives, the search was generally associated with a white, Protestant middle class, favoring public over private interests in an attempt to contain corruption and other outgrowths of an unchecked capitalism. Whereas Richard Hofstadter's classic *The Age of Reform* identified "status anxiety"—the fear of the established elites of losing control—as the core of this quest, Wiebe described the more or less successful pursuit of modern citizens to make society more rational and efficient with the help of an expanded state. Gabriel Kolko, in his study *The Triumph of Conservatism*, and other scholars of the New Left argued that the reformers' intent to curb the power of big business through regulation never resulted in truly transformative measures while the spheres of politics and economics in fact grew closer together, enabling the representatives of the industrial system to tighten their grip on American society. In her pathbreaking study *Manliness and Civilization*, Gail Bederman delineated the role of shifting gender and racial identities, arguing that male middleclass Americans were responding to perceived threats posed by assertive workers, immigrants, African Americans, and the "new women." This peek at the historiography already demonstrates that the search for order was a zigzagging path between fear and optimism, modernism and reaction, articulating both the wish for harmony and the endorsement of aggression—a social and cultural struggle over meanings, power, and control.

The building of the Panama Canal, materializing in the midst of a foreign, tropical environment, was one of the wonders of the age that "haunt and lure the imagination," as Lippmann had put it, and a symbol of the new American empire. For most people, it could not be experienced firsthand (unless they traveled to the Isthmus, as many tourists did) but had to be imagined and interpreted through texts and images. This was the task of the Panama authors. Most of them were journalists and writers addressing the American middle class, a large social group that was undergoing far-reaching changes itself. As Warren Susman observed, historians are "of course always inventing new middle classes."[41] It is difficult, and perhaps futile, to define the audience these authors were addressing, but there are some guideposts. Most of its members were employees, not workers or entrepreneurs, spanning the entire career spectrum from clerks to managers. Many of them worked for large corporations that were slowly replacing the machine shops of the early industrial era, offering an increasing amount of office jobs, and many were members of what we now call the tertiary or service sector, a labor segment profiting from the emerging culture of consumption that was associated with jobs in sales, marketing, and advertising. Others were "professionals," such as lawyers and doctors, who, like the administrators and engineers in government and big business, increasingly relied on specialized knowledge and expertise. They all shared an interest in the world around them, hoping to keep the transformations of the modern era under control and create a new (albeit illusive) civic unity. They joined organizations and listened to politicians. These "challenged" middle-class Americans spearheaded the search for order. Last but not least, they had the money and the leisure time to spend on books and magazines.

The evolution of the media played an important role in the forming of this new middle class.[42] Whereas most newspapers in the United States catered to a local audience and expressed distinct political preferences—and continue to do so even in the twenty-first century—the magazines addressed a national audience, joining forces with advertisers who hoped to reach large groups of people with above-average income and purchasing power. Mass production, the consolidation of industry, resulting in the rise of corporations, and professional marketing instruments had helped to create products for a "national" consumer. Before the invention of radio and television, magazines were the first medium to exploit this new business opportunity. From 1890 to 1905, the number of national magazines almost doubled and circulation tripled, and they continued to gain new readers.[43] The subscription numbers of the *National Geographic Magazine*, which brought the wonders of foreign places and people into the American home, soared from 3,400 in 1905 to 107,000 in 1912.[44] Advances in print technology made the use of images cheaper and easier. Established magazines such as *Atlantic Monthly, Scribner's*, or *Century* were challenged by cheap, mass-circulation competitors such as *Cosmopolitan* or *McClure's*, which used journalistic rather than essayistic formats, and they adapted to the new trend. The magazines reported on the latest fads in the big cities, on social and technological progress, on world's fairs, and, of course, on the rise of the American empire.

Many Panama authors were among the writers for these magazines, and whether they were professional journalists, were administrators who jotted down their memories of the Canal building, or were in some other respect involved with the project, they all were engaged in the search for order. Their accounts of the work and life on the Isthmus appeared in magazines and books. It can be assumed that the full-length popular histories, despite their lack of advertising, were an economic success. Willis John Abbot's *The Panama Canal in Picture and Prose*, for instance, sold more than one million copies.[45] Predictably, public attention culminated shortly before the completion of the seaway. Sometimes, books written during an earlier phase of construction were republished under a different name: While Arthur Bullard's *Panama: The Canal, the Country and the People* (1914) appeared under his real name as opposed to the pseudonym Albert Edwards used in the edition of 1911,[46] Charles Forbes-Lindsay's *The Isthmus and the Canal* (1906) was published in a slightly extended version (and with an identical inscription!) as *The Story of the Panama Canal* in 1913, by Logan Marshall. Both were real persons who had written other books. Frequently, chapters of the Panama authors' works had been published before in magazines, in Bullard's case in *The Outlook* and *Harper's Weekly*. Abbot and Bullard as well as Ira Bennett, Frederic Haskin, Willis Johnson, and Hugh Weir were journalists; some of them focused on travel and foreign reporting, others wrote novels and self-help books. Forbes-Lindsay, for instance, had published an introduction to the card game bridge; William Scott, author of *The Americans in Panama*, also wrote a classic polemic on tipping, *The Itching Palm*.[47] These examples demonstrate that the Panama authors belonged to a group that the historian George Cotkin has called the "cultural custodians." They were translators and interpreters of social change.[48] In some cases, the authors would stress the authenticity of their descriptions, adding clauses such as "approved by leading officials connected with the great enterprise" as if their reporting alone lacked credibility.[49] Photos and other images of the Canal played a crucial role in these books and articles. Many authors had spent only a few weeks or even days on the Isthmus, and the subject of their books was indeed an "imagined territory,"[50] a reference to the promises and perils of progress rather than to an actual place. Their works interpreted the Canal construction and the American state on the Isthmus as examples of a new public spirit—a national effort to contain confusion and chaos.

Another group of Panama authors was or had been directly involved in the Canal project, whether as a census taker like Harry A. Franck, a traveler and lecturer who wrote *Zone Policeman 88,* as a stenographer, like Mary Chatfield, author of *Light on Dark Places at Panama,* or as a Canal Commission executive, like Walter Pepperman with his account *Who Built the Panama Canal?* The person who was most closely associated with the "official" work on the Isthmus was Joseph Bucklin Bishop, the secretary of the third Canal Commission. Bishop had worked as a newspaper writer and editor in New York, where he had met Theodore Roosevelt, then a commissioner in the city. In 1905 Roosevelt appointed Bishop as a kind of public relations officer for the Canal project, based at first in Washington, D.C., and from 1907 on in Panama. Bishop stayed on the Isthmus through the entire building period and was a crucial factor in the interpretation of the Canal for the American public, both as a writer and as a source and facilitator to other authors, editors, and officials in politics and business. After 1914, he returned to his old profession and became Roosevelt's biographer.[51] On the Isthmus, he served as the "ghost writer, policy adviser, alter ego" of Chief Engineer George W. Goethals.[52] Bishop also edited the *Canal Record,* a weekly newspaper distributed to the labor force as well as to journalists in the United States.

Accompanying Bishop's letter of appointment in 1905 was a memorandum by the chairman of the Canal Commission, stating that Bishop would "have charge of the publicity and literary branch of its work." The note continued: "He will also be the official historian of the Canal, preserving and compiling the authentic and authoritative record of its construction."[53] This assignment resulted in Bishop's book *The Panama Gateway,* one of the most significant documents published by the Panama authors. (His son Farnham, author of *Panama Past and Present,* joined their ranks as well.) Bishop's background and his role on the Isthmus illustrate how closely the official views on the Canal construction and the interpretations of writers employed by magazines or publishing houses were linked to one another. For access to and material on the Canal work, especially photographs, the authors were dependent on the government. On the other hand, officials like Bishop wrote books similar in content and style to those composed by independent journalists. The government was keenly aware of the fact that the building of the Panama Canal needed to be explained to a national audience. To prevent "false and misleading" reporting on the Canal, Bishop had advised his superior, "the publication of interesting and authentic information" would be necessary.[54] As a consequence, the authors wrote, showed, and sometimes (as in the case of Theodore Roosevelt) even embodied the success story of the Canal. And yet, in their search for order, fears of failure and of foreign people and places both at home and abroad can be sensed, often suppressed by a diffuse optimism and the vision of a collectivist but segregated society. Perhaps nothing better illustrates the yearning for a utopian middle-class community than the interpretations of the Panama Canal.

Culture and Discourse: From Texts to Reality

Policymakers, intellectuals, and journalists, including the Panama authors, were constructing the Panama Canal through their interpretations. The study of this process is what I mean by cultural history. In the 1970s, before the onslaught of poststructuralist theories, the ethnographer and theorist Clifford Geertz applied a corresponding concept to his own discipline: "Culture, here, is not cults and customs, but the structures of meaning through which men give shape to their experience; and politics is not coups and constitutions, but one of the principal arenas in which such structures publicly unfold."[55] The task of the ethnographer/historian is to study the shaping and unfolding of "webs of significance."[56] In the course of the cultural turn, scholars in all fields began to study categories previously viewed as fixed (or even biologically determined), such as gender or race, as social constructions, as *texts* that could be inscribed into individual or collective bodies. A concept such as the nation, as Benedict Anderson has pointed out, refers to an "imagined community" of meanings and identities, constituted by a specific group during a certain time period, rather than to an enclosed space or a common legacy.[57] Cultural analysis, in the Geertzian sense, is an interpretive effort—Geertz says science, which I find too difficult to define in the context of the humanities—in search of meaning.[58]

The phrasing of this definition implies that the practice of the historian engaged in cultural analysis may not be so different from what the "sources" were doing. Every interpretation only adds another layer to the interpretations already made. If the alleged "reality" of history is to be found at the bottom of this pile, then whose reality would this be? "Right down at the factual base, the hard rock, insofar as there is any, of the whole enterprise, we are already explicating: and worse, explicating explications," Geertz notes.[59] In this light, Leopold von Ranke's famous definition of the historian's task—to find out "wie es eigentlich gewesen" (how it has actually been)—becomes a futile and literally "meaningless" endeavor obscuring "the endless character of the interpretive process."[60] Historical writing, as any practitioner would confirm from experience, is essentially an exercise in intertextuality.[61] Like the anthropological scholarship to which Geertz was referring, it is not false or unfactual by virtue of its limitations but rather "'something made,' 'something fashioned'" from the perspective of its author and his own time and place.[62]

I try to acknowledge this (usually concealed) circumstance by pointing out that I have written this book not only as a white, Protestant male—attributes I share with many of the actors in this study—but also as a German citizen, a foreign scholar of U.S. history, and a professional journalist. In some instances, this information may help the reader to better understand my approach to the subject. I chose the structure of this study not only for purposes of readability and the unfolding of my argument but also in partial imitation of the books written by the Panama authors. In many ways, my narrative resembles their popular accounts of the Panama Canal building, in which the prehistory of the waterway, the actual construction work, the Canal Zone state, and the Panama-Pacific International Exposition were often discussed in a similar thematic and chronological order. This setup may remind the reader that the historian, despite all attempts to achieve something new, is always building his or her interpretations on somebody else's.

A concept that is also crucial to this study and needs explanation is described by the term "discourse." For more than three decades, it has played an important role in the evolution of the humanities, beginning with the work of the French philosopher Michel Foucault. Even if it has not yet become part of the broad mainstream of historical scholarship, discourse analysis, as mentioned earlier, is intrinsically linked to the linguistic turn. It takes place, as Alun Munslow writes, on a "language terrain,"[63] assigning a new significance to the study (and definition) of texts. The practice of discourse analysis is diverse, dynamic, and often, as in this study, patchwork. Many theorists have contributed to its development, and in light of the wealth of their material, I do not venture to base this book on any single approach or even the work of a single scholar. There is a great risk, therefore, of methodological inconsistency and of succumbing to an unfocused, consensus view of discursive processes, but I am willing to take this risk.

A discourse refers to a certain formation and configuration of language. It is perhaps confusing to speak of texts as the "matter" of this configuration. Instead, the notion of texts should be expanded to other expressions of a textual (and thus interpretive and constructed) nature, such as images and practices.[64] Geertz, for example, discusses gestures and other forms of physical human behavior pertinent to ethnography. For the historian, many of these interpretive expressions are not available as sources, but the growing body of scholarship on images and visual cultures acknowledges the need to extend the analysis beyond the written word. A discourse, as I would define it, is a relational field of text and context or, very simply put, the way a certain story can be told, the articulation or communication of a certain subject, with the text (in the expanded definition) as its instrument.[65] It necessarily operates synecdochically—as a part for the whole (and vice versa)—through representations, metaphors, and symbols. It produces sociocultural definitions, rules, and meanings for the recipients and therefore not only reflects reality but actually constitutes it by determining or at least indicating what can be said, thought, and experienced with regard to a certain subject (and, by reciprocity, which texts can then be created about it). These texts and meanings are no longer "free" or even arbitrary.[66] The discourse is, in this sense, *material*—there is no distinction between "ideas" and "reality."

An important question arises: What actually drives this process? Who is behind it? It helps to borrow an analogy from the physical sciences, if only for illustrative purposes and with no claim to correctness. Like an electric field, the relational field of the discourse exists due to an elementary force, revealed in a human "voltage," a differential of power: the search for order, meaning, and identity. This need for control—not necessarily in a repressive but in a "productive" way, expressed in the production of "truths" and meanings[67]—is the reason why discourses exist. It is crucial in this respect to point out—having left behind the electric field analogy by now—the contested nature of discourses. Imbalance creates contradictions and discontinuity. Although "the existence of a dominant synecdochic integrative thought"[68] is constitutive for the discourse, it has fissures, too.[69] This is precisely what makes discourse so valuable as a methodological concept and so potent in practice. It covers more ground than a single idea or ideology; it strengthens compliance but also invites resistance.

From a strictly Foucauldian perspective, the authors of discursive texts are mere labels, and their subjective intentions bear no significance for the analysis. The historic, "de-authorized" texts are part of a system called archive, and the practice of uncovering its structure is called archaeology.[70] While I still acknowledge the authors of texts as individuals, my approach remains fundamentally different from traditional intellectual history. The Panama authors did not merely articulate "ideas" about the Canal. They regulated what could be thought and believed about it—and about American society. For their middle-class audience, the writers' construction of the Canal was more relevant than the concrete on the Isthmus (even if it would enjoy a much shorter life span). I discuss authors such as Joseph Bucklin Bishop and Willis John Abbot with regard to their discursive and not their personal impact on the "construction" of the Panama Canal. This distinction also applies to Theodore Roosevelt, who plays only a minor role in this study as a political actor. I am much more interested in his significance as an interpreter of manliness in the imagining of the Canal.

There is a school of criticism that accuses discourse analysis of reductionism: History becomes a subdiscipline of literary studies, or worse, literary theory. Texts, it is claimed, are less real than the mud dug up from Culebra Cut in Panama, or the beating a West Indian laborer takes from his American foreman. Political institutions as well as economic and social mechanisms are deemed outside of the focus of the interpretive effort. While the Canal lives on, many of the interpretations and meanings described here did not survive World War I—which, incidentally, may be one of the reasons why they have not commanded the attention of historians. There is no other way to settle this argument than to point out, like Norbert Finzsch has done with respect to the brutal killing of the African American James Irvin by a white mob, that even such ordeals, as representations, often resemble texts—apart from the obvious fact that they may only be accessible for the historian through texts—while acknowledging that, of course, there is also a nondiscursive reality.[71]

In this study, I regard the texts and images employed by policymakers, intellectuals, writers, and artists as constituents of a discourse revolving around the Panama Canal. The subject of its synecdochical pull is the search for order within the American middle class, the interpretation of changes caused by technological advances, territorial expansion, and social upheaval. This is an alternate history of the construction of the seaway, which takes place not on the Isthmus but within the United States—not in excavations of dirt but in people's minds. The Panama Canal was an astonishingly integrative device for interpretation, easily surmounting the borders of the foreign and the domestic, of diplomatic and cultural history. It can serve as a new lens for the study of the U.S. entry into the modern age.

Chapter 1 discusses the previous histories of a Panama Canal, including the failed French effort, the far-reaching involvement of the United States in the region as well as the events resulting in the creation of Panama as a "sovereign" country and the eventual commencement of the work on an American waterway. It places special emphasis on the social visions of the expansionist intellectuals Alfred Thayer Mahan, Brooks Adams, Frederick Jackson Turner, and, most important of all, Theodore Roosevelt. When the historian and self-styled cowboy became president, he made the construction of the Panama Canal the center of his program to build a manlier nation. Perceptions of gender and race were inscribed into the Canal project from its start. When progress was slow and critics of the Canal sensed failure, Roosevelt took his trip to the Isthmus and later installed a new administration, paving the ground for the success stories of the Panama authors.

In chapter 2 I examine the authors' attempts to interpret the Canal itself. After an excursion on the containment of disease in the Canal Zone and its role in the popular writings, I turn to the writers' constructions of the waterway as a civic symbol similar to and yet different from earlier public works of technology. Both passage and network, the Canal evoked seemingly opposite concepts: past and future, war and peace, the national and the international. I discuss the attempts of the authors to make the engineering feat believable and meaningful. Their interpretations of the French failure on the Isthmus served as a mirror and a call for applying the lessons of Panama to an American society perceived as morally corrupt and lacking efficiency and resolve.

The images of the Panama Canal and their use by the authors are discussed in chapter 3. As most of the photographs were provided by the government, I examine the context of their creation, especially the role of the "official photographer" on the Isthmus. For this analysis, I rely extensively on administrative records and letters held at the National Archives. I also address the use of now obsolete media such as stereographs and the visual representation of Roosevelt's visit to the Canal Zone. Although neglected by the previous scholarship, images were part of the Panama discourse; they were powerful, constructed tools of interpretation, of explaining and charting the new American empire.

Chapter 4 turns to the Panama authors' interpretations of the administration and daily life in the Canal Zone. Ignoring the majority of West Indian workers, the writers encountered a collectivist mini-state ruled and inhabited by white Americans, at its head a "benevolent despot," Chief Engineer Goethals. This suburban society on the Isthmus was reminiscent of Edward Bellamy's utopian vision of the year 2000 and became the blueprint for a future American nation. The authors' accounts were closely linked to the domestic discourses on reform, and I try to shed light on this relationship by considering another influential group of writers, the New Intellectuals, and their views of the Canal effort. This chapter also addresses the impact of World War I, whose outbreak coincided with the opening of the Panama Canal, on the social visions associated with the project.

At the Panama-Pacific International Exposition in San Francisco, the stage was set for an alternate construction of the Canal's meanings—from the perspective of the nation and the "race" as a whole but also with respect to the rising California city that had barely survived an earthquake nine years earlier. In chapter 5 I discuss the images and exhibits of the fair, covering the architecture and sculpture as well as the gigantic model of the seaway in the exposition's midway "Zone." The fair was another complex text revolving around the Panama Canal and the American middle class, extolling the achievements and expansion of a white, male civilization while also testifying to the underlying fears of change.

1

~

Logistics of Expansion

The Long Road to Realization

From Columbus to the French Debacle:
Prehistories of the Canal

For the European explorers and conquerors, the Isthmus of Panama was a historic place. On his fourth voyage, Christopher Columbus entered the mouth of the Chagres, still believing he was facing the coast of East Asia. Nine years later, in 1513, the Spaniard Vasco Nuñez de Balboa allegedly became the first European to see the Pacific Ocean—in the vicinity of today's Panama City. In 1520 Fernão de Magalhães (Ferdinand Magellan) sailed through the strait named after him at the southern tip of Latin America, establishing the sea route most ships would take in the centuries prior to the building of an interoceanic canal.[1]

The idea of a waterway across the continent in Panama or other suitable locations in Nicaragua and Mexico (through the Isthmus of Tehuantepec) intrigued statesmen and scientists long before it became technologically feasible. Habsburg emperor Charles V commissioned geographical studies, while his successor Philip II, discouraged by the results and convinced that building a canal would constitute an illegitimate interference with God's creation, abandoned the plans. Enthusiasm surged again at the beginning of the nineteenth century. In 1811 the German explorer Alexander von Humboldt presented the first quasi-scientific study of possible canal routes, even if it relied more on hearsay than established facts. Humboldt shared his interest in geography with Thomas Jefferson and had been the guest of the American president in 1804. In his *Political Essay on the Kingdom of New Spain* (1811), he wrote that an interoceanic canal was "of the greatest interest for the *balance of commerce,* and the political preponderancy of nations."[2]

The collapse of the Spanish colonial empire put the young American nation in a new position. The Monroe Doctrine, proclaimed in 1823 by President James Monroe, stated that all of Latin America pertained to the interests of the United States and that, in the long run, no European power would be tolerated in the hemisphere. Panama became one of the first manifestations of the doctrine. Since its independence from Spain in 1821, the region had been part of the dominion of Colombia. In 1846 the Bidlack–Mallarino Treaty was signed, granting the United States free transport and trade across the Isthmus of Panama, a land passage that had been in use since the sixteenth century. In return, the United States would guarantee Colombia's sovereignty and send troops if the transit across the land or through a potential canal was endangered. Until the independence of Panama in 1903, the American government launched fourteen military interventions under this treaty—even while the French were building a Canal there.[3] Following the Mexican-American War in 1848, the territories of California, New Mexico, and Texas fell to the United States. From the same year on, the gold rush in California drove thousands of "pioneers" across the Isthmus. It became an important passage from the East to the West Coast, for purposes of trade as well as travel.

In contrast to the Bidlack–Mallarino Treaty, the agreement reached with Britain in 1850 displayed little of the spirit of the Monroe Doctrine. In the Clayton–Bulwer Treaty, the two powers consented that neither of them would have exclusive control over an interoceanic canal in Central America and ruled out its military use by fortification. For more than fifty years, this agreement would prove a hindrance to American ambitions in the region.

The gold rush brought many ambitious Americans to Panama. In search of a good fortune in California, they took the risk of the dangerous passage from the Caribbean port of Colon to Panama City on the Pacific side, in spite of tropical diseases and frequent assaults by thugs. New York businessmen soon realized that a railroad connecting the two ports would generate good profits. Laborers from the West Indies,

China, Africa, and Ireland arrived at the Isthmus to work on the toilsome project. Thousands of them died of illnesses; later, historians reported (and dismissed) the legend that for every track of the railroad, a laborer had given his life. In 1855 the first "transcontinental" railroad—fourteen years before the route across the North American continent was completed—opened and became an instant success. Paying twenty-five dollars in gold for the passage, more than 400,000 passengers crossed the Isthmus by train during the first decade after the opening. By virtue of the Bidlack–Mallarino Treaty and the railroad enterprise, historian Michael Conniff argues, Panama had already become a "protectorate of the U.S. government."[4] During the Civil War, President Abraham Lincoln even considered the colonization of southern blacks in Panama.

Despite the close U.S. ties with Panama, it was still far from clear where an Isthmian Canal could and would be built. Treaties were signed with Nicaragua and Colombia in 1867 and 1869 to ensure transit across a future canal, but the latter treaty was never ratified. Few details were known about the topography of potential canal routes. Beginning in 1870 with the Darien Expedition to the Isthmus of Panama, headed by Navy Commander Thomas Oliver Selfridge Jr., the president's Interoceanic Canal Commission carried out seven expeditions to the region and eventually settled on the Nicaragua route. Meanwhile, a French syndicate managed to secure a concession from Colombia for the construction and neutral operation of a Panama Canal by a private company for ninety-nine years. Because of the advances in hydraulic engineering, the project imagined for centuries now actually seemed realizable. Against all odds, the former French diplomat Ferdinand de Lesseps (1805–1894) had built the Suez Canal from 1859 to 1869. An aged national hero and an excellent promoter set to crown his life's work with yet another epochal engineering feat, de Lesseps bought out the syndicate with the help of other investors, set up the Compagnie Universelle du Canal Interocéanique, and also acquired the stocks of the Panama Railroad. From a political standpoint, there was little the United States could do to prevent the private project from realization. All of a sudden, the interoceanic waterway was about to materialize not as an American or British canal but as a French canal.

De Lesseps raised 300 million francs ($60 million) in a public stock offering. It was "public" in the best sense since around eighty thousand stockholders had bought fewer than five shares each. And it was twice oversubscribed—de Lesseps could and should have raised much more money. Once again, thousands of foreign workers toiled in Panama, but progress was slow. Yellow fever and malaria took their toll; the exact numbers are controversial, but the chief sanitary officer of the American Canal force estimated in hindsight that up to one-third of the white laborers died on the Isthmus.[5] The crucial mistake was de Lesseps's conviction not to include locks but to build a sea-level canal as he had done in Egypt. When the Compagnie's equity and debt of more than 1 billion francs was spent, de Lesseps tried to get the French government's support for a lottery bond worth another 600 million francs. He even agreed to abandon the sea-level concept and in 1887 hired engineer Alexandre Gustave Eiffel, who was building the Eiffel Tower for the World's Fair to be held in Paris two years later, to design and build the locks for the canal. But there was not enough interest in the bond, and in 1889 the Compagnie went bankrupt.

In France the failure of the Canal project resulted in a national crisis. The anti-Semitic journalist Edouard Drumont charged its leaders and supporters with corruption, triggering an official investigation that ruined the reputation of de Lesseps and uncovered government involvement at the highest levels.[6] Two ministers of the French cabinet had to step down. The press, it turned out, had continued to cheer for the enterprise despite evidence of its financial difficulties—and accepted money for the enthusiastic coverage. The ailing de Lesseps, a broken hero, died in 1894. "For the French, Panama, the fabled door to the riches and wonders of the Pacific, became a sign of all that was wrong with the republic," historian Matt Matsuda writes.[7] The exclamation "quel Panama!" was used as a synonym for scandal and ruin. Drumont had intended to portray Jewish financiers as the scapegoats for the multifaceted failure of the project. His anti-Semitism proved popular, and it carried on into the so-called Dreyfus affair. In 1894 Captain Alfred Dreyfus, a Jewish officer in the French army, was arrested, convicted for treason, and deported. The debate over the legitimacy of the charges and the motives behind them divided the country for years, and France plunged into yet another crisis. Meanwhile, on the Isthmus, a new French Canal company managed to secure an extension of the concession, hoping for a buyer of the infrastructure. Excavations in Panama continued on a minimum level.[8]

After the French debacle, there was little doubt that the United States, a rising economic superpower, would eventually build the canal.

In the Hay–Paunceforte Treaty, the country finally managed to reach a clear-cut settlement with Britain. The European power, suffering from the Boer Wars in South Africa, agreed to surrender the shared control it held on paper on a future canal in the Caribbean. In the first version of the treaty, the United States did not receive permission to fortify the seaway. In the second draft, this obstacle was removed. In early 1902, boosted by growing political support at home, the Canal project was almost on its way. There was a general consensus that it would be built in Nicaragua, the option favored by John Tyler Morgan, chairman of the influential Senate committee on interoceanic canals.

Within a few months, the "Panama lobby" reversed the situation. Its protagonists were William Nelson Cromwell, a lawyer from New York and representative of the new Canal company, and Philippe Bunau-Varilla, who had served as chief engineer under de Lesseps at the age of twenty-six. The cunning Frenchman focused all his energies on the completion of the Canal in Panama and even tried to convince the Russian czar to take over the project. The arguments in favor of Panama were not difficult to grasp: the country had established ports and a railroad for transport, the territory was fairly well known, and the Nicaragua route would require more locks and more digging. The French concession including rights and equipment was on sale for $40 million. Slowly, experts and influential magazines such as the *Scientific American* started to endorse the Panama route.[9] Although President Theodore Roosevelt, who wanted to close the deal before his reelection campaign in 1904, had become a staunch supporter, members of Congress were still divided on the issue.

Aided by a psychological trick, the lobbying efforts paid off. Even though the Nicaraguan government had told members of the Senate that a volcano threatening to endanger a prospective canal was dead, an official stamp showed the mountain of Monotombo spitting lava. Shortly before the decisive vote, Bunau-Varilla bought every single one of these stamps in stock in Washington, D.C., and placed them on the senators' desks. The anecdote probably reveals more about Bunau-Varilla's personality than anything else, but nevertheless the Senate decided in favor of the Panama route on June 19, 1902—by a small margin of 42 to 34. The legal details of the construction were settled in the Spooner Act, which became law on June 28. The Canal would be governed by an Isthmian Canal Commission reporting to Secretary of War William Howard Taft.

Unexpectedly, the final hurdle proved the most difficult. In the Hay–Herrán Treaty, the United States and Colombia negotiated a one-hundred-year lease on a six-mile-wide Canal Zone, excluding the major port cities. Colombia remained the sovereign, but the United States was allowed to set up its own jurisdiction in the zone. In return, it paid Colombia a lump sum of $10 million and an annual rent of $250,000. The treaty was signed on January 22, 1903, but the Colombian parliament refused to ratify it unless the country was granted a larger sum as down payment. The Canal deal was stuck.

Laying the Ground: The Expansionists' Visions

When the building of an interoceanic canal under the auspices of the American government seemed imminent, U.S. intervention in the affairs of other countries or regions was already an established process.[10] The most widely accepted rationale for U.S. expansion abroad was economic. To policymakers and businessmen, the depression years of 1893 to 1897 indicated that the country needed access to foreign markets, preferably in Latin America and Asia, to sell surplus goods coming out of American industry and agriculture. Overproduction—the main issue at the National Association of Manufacturers' meeting in 1895—threatened to result in the self-destruction of the economic system, expressed in strikes and other demonstrations of social unrest.

There was a consensus that the United States, itself a former British colony, would not strive for a European-style colonial empire, but the alternative format of American expansionism remained controversial and, in practice, ambivalent. The policies carried out in the 1890s and 1900s ranged from "soft" measures such as trade agreements to disputes with European powers and outright annexation in order to secure strategic military bases in other countries—an instrument firmly opposed by the so-called anti-imperialists. Within only a few years, the American involvement abroad had reached large dimensions. Reciprocity agreements with Latin American countries, setting up mutually reduced tariffs for exports, were followed by, among other events, the annexation of Hawaii—pending since 1893 and finally enacted in 1898 after decades of economic dependency—the settlement of a border dispute between Venezuela and British-controlled Guiana in 1896, and the joint acquisition of the Pacific island group of Samoa with Britain and Germany in 1899. These actions were complemented by the diplomatic approach of the Open Door: named after the notes sent by Secretary of State John Hay to European governments in 1899 asking them not to violate international trading rights in China, this policy reflected more subtle yet assertive attempts to ensure and expand American access to foreign markets.

The most consequential conflict of the decade emerged in 1898 with Spain. A Native revolt against the colonial power in Cuba enjoyed public support in the United States, and when the U.S. battleship *Maine* exploded in the port of Havana—by accident, as the evidence would show later—President William McKinley, under pressure by the warmongering yellow press, asked Congress for a declaration of war. A few months later, in the Treaty of Paris, Spain acknowledged Cuba's independence and ceded the Philippines, Puerto Rico, and the Pacific island of Guam to the United States. Hay called the quick and largely painless conflict "a splendid little war." Cuba stayed under official U.S. control until 1902, with military bases such as Guantánamo Bay, today a high-security prison for alleged terrorists, remaining there until the present. In the Philippines, the United States acquired a full-fledged colony for the price of $20 million and would soon become entangled in a bloody guerilla war. There could be no doubt any more that the country had turned into an empire.

Economic motives were a crucial factor for expansion and, by its very nature as a trading route, the building of a Panama Canal. Historians have pointed out, however, that a strictly mercantilist view does not suffice as an explanation for U.S. policies abroad and their support at home around the end of the nineteenth century. A group of influential intellectuals and policymakers whom I call "the expansionists" pursued an agenda incorporating economic impulses into a larger scheme of U.S. interests, grounded partly in their negative perception of industrialization and its effects on American society. Robert Wiebe argues for a distinction between "profit-oriented" and "power-oriented" actors,[11] and J. Michael Hogan writes about the "internationalists" who "promoted an ideology of strategic superiority and national spiritual health."[12] To shed more light on these rather vague terms, the ideas of Alfred Thayer Mahan, Brooks Adams, Frederick Jackson Turner, and Theodore Roosevelt are explored in some depth in this and the following section. Their gendered and racialized views were the basis for and therefore part of the Panama Canal discourse unfolding in the writings

of the storytellers during the building period. In response to the implications of social and technological change, the expansionists constructed the vision of a future America, and the Panama Canal was at
the center of this vision.

Apart from its significance for merchant vessels, the building of an
interoceanic canal was attached to the issue of American sea power. In
1880 the ships of the American navy had antiquarian value at best. Out
of 1,942 vessels, only forty-eight were able to fire shots at an opponent,
and for many years, the competitiveness of the U.S. Navy in comparison
to European sea powers such as Britain and, more recently, Germany,
had not been an urgent public issue.[13] This assessment changed with the
publication of Captain Alfred Thayer Mahan's (1849–1914) study *The Influence of Sea Power upon History* in 1890.[14] Although the naval strategist
and historian examined only the time period from 1660 to 1783, he managed to convey a sense of the impact of sea power on geopolitical developments and provided a quasi-scientific argument for the buildup of a
new American navy. Mahan gained support from like-minded spirits
such as Theodore Roosevelt, himself a historian, who stated in the *Atlantic Monthly* that Mahan had written "distinctively the best and most important, and also by far the most interesting, book on naval history which
has been produced on either side of the water for many a long year."[15]

In his essays of the following years, Mahan argued for the building
of a modern navy. A commercial fleet transporting American goods to
other countries was useless unless protected by battleships. As a consequence of motorization, the oceans were transfigured into a new, modern space, "a great highway," whose control promised "predominant influence in the world."[16] The interoceanic canal was a crucial element in
Mahan's vision, a node in a web of power. What Columbus had searched
for with "a seer's eye" in the Caribbean was now the key to dominance:
"The secret of the strait is still the problem and reproach of mankind."[17]
Mahan employed the presumed public support of his argument to silence his critics: "Whether they will or not, Americans must now begin
to look outward. The growing production of the country demands it.
An increasing volume of public sentiment demands it. The position
of the United States, between the two Old Worlds and the two great
oceans, makes the same claim, which will soon be strengthened by the
creation of the new link joining the Atlantic and Pacific."[18] During the
1890s, the construction of modern battleships finally commenced, and
the new navy was put to the test during the Spanish-American War. The

expansionists criticized that it took the *Oregon,* anchored on the West Coast, almost two months to reach the scene of war in the Caribbean after sailing around Cape Horn. This argument would be repeated over and over again by the Panama authors to demonstrate how urgent the construction of a canal appeared to be at the time.[19]

For Mahan, the quest for sea power had an external as well as an internal function. Control of the oceans enabled the Western nations to spread their common legacy around the world. "Wherever situated, whether at Panama or Nicaragua, the fundamental meaning of the canal will be that it advances by thousands of miles the frontiers of European civilization in general, and of the United States in particular, that it knits together the whole system of American states enjoying that civilization as in no other way they can be bound," he wrote.[20] As opposed to old-style warriors in Europe, Mahan did not envision a war between the powers to sort out who was the strongest, but a kind of international peace order to preserve and defend the economic and racial supremacy of the West. Military "preparedness," one of Mahan's watchwords, was a national goal in itself, a modern state of power. He wrote that "public conviction is a very different thing from popular impression, differing by all that separates a rational process, resulting in manly resolve, from a weakly sentiment that finds occasional hysterical utterance."[21] For Mahan, expansion was also a moral weapon against the insecurities within society.

The historian Brooks Adams (1848–1927), brother of the author Henry Adams, shared this belief. In 1896 he had written *The Law of Civilization and Decay,* trying to extract an understanding of the present time from patterns of history, an approach very similar to Mahan's analysis of sea power. The book was well received in Washington. Adams admired the imaginative spirit of the Middle Ages, symbolized by the Gothic cathedrals of Northern France. He theorized that history went through cycles of increasing centralization, only to disintegrate again "because the energy of the race has been exhausted."[22] For him, the "rise of the bankers" and "advance of cheap labor"—both representative of the social crisis of the present capitalist system in the U.S.—indicated that such a collapse was imminent.[23] Searching for the "power necessary for renewed concentration," Adams turned to expansion.[24] In his follow-up work, *The New Empire* (1902), he argued for American dominance of Asia, carried out by a charismatic leader such as Theodore Roosevelt and overseen by an efficient modern state protected against the pitfalls of an unrestrained capitalist order. The United States was an "advancing social cyclone," he wrote, urging for administrative reform and a military spirit.[25] Like Mahan, Adams assigned a moral function to expansion. The challenges facing industrial America would not be solved by extending the exhausted economic system to other countries, but rather by overcoming this system, transforming it into a rational, collective, and "manly" social organization of global dimensions.

Adams and other expansionists were heavily influenced by evolutionary theory and especially its application to economic and social problems, as expressed in the work of the British philosopher Herbert Spencer and his terminology of the "survival of the fittest." Summarized under the term "social Darwinism," these theories were used to legitimize the capitalist system of competition among businesses and individuals as well as the struggle between the different peoples and nations of the world for power and dominance.[26] For the white elites in the United States, there was little doubt that their "race" would prevail. Civilization was interpreted as a "precise stage in human racial evolution" following savagery and barbarism.[27] At the time, the belief that other ethnic groups, among them Native Americans, African Americans, and immigrants from Southern and Eastern Europe, were inherently inferior was "endemic."[28] This racist view was supported by biological theories that would only slowly be disputed, for instance by the work of the anthropologist Franz Boas in the 1910s.

Racial superiority was also granted to Britain, the former colonizer. In 1885 the public orator John Fiske had popularized the idea of a global Anglo-American federation. His *Manifest Destiny* lecture went through multiple reprints and proved as influential as *Our Country* (1885), the main work of the preacher Josiah Strong, who called for an Anglo-Protestant "mission" to the backward peoples of the world.[29] "Anglo-Saxonism," as these postulations of supremacy were named, was enthusiastically endorsed by the expansionists: "Among all foreign states, it is especially to be hoped that each passing year may render more cordial the relations between ourselves and the great nation from whose loins we sprang. The radical identity of spirit which underlies our superficial differences of polity surely will draw us closer together, if we do not set our faces wilfully against a tendency which would give our race the predominance over the seas of the world," Mahan wrote.[30]

The belief in racial supremacy was an ambivalent expression of expansionism. The missionary aspiration to assume responsibility for others was accompanied and, arguably, outweighed by the search for a national identity and moral self-assurance. The focus of Adams's *Law of Civilization and Decay* was the evolution of Western society, resulting in the present stage of crisis in the United States. While his "law," not unlike the manifestos of Karl Marx and other socialist thinkers, was in fact a deterministic model of history, he still believed that it could be repealed. Instead of letting the evolutionary mechanism proceed fatalistically, there was the option of applying it deliberately: American society needed to improve and reassert itself. The theoretical basis for this process was not to be found in the work of Spencer but his French colleague Jean-Baptiste Lamarck, whose ideas had little enduring scientific value but proved suitable for the kind of moral transformation the expansionists were proposing. Lamarck's principle of "acquired characteristics" stated that new ways of behavior enabling animals and humans to survive could be inherited. Projected unto a group of people, this neo-Lamarckian interpretation implied that the common will of a nation was determining its chances of survival and domination.[31] In his depiction of sea power, Mahan elaborated this assumption: "The greatest of the prizes for which nations contend, it too will serve, like other conflicting interests, to keep alive that temper of stern purpose and strenuous emulation which is the salt of the society of civilized states, whose unity is to be found, not in a flat identity of conditions—the ideal of socialism—but in a common standard of moral and intellectual ideas."[32]

The complex rationale of expansionism was perhaps most concisely formulated in Frederick Jackson Turner's (1861–1932) *frontier thesis.* The historian had first put forward his ideas on the presumed end of westward expansion in a lecture to other members of his profession at the World's Fair in Chicago in 1893. After the publication of *The Significance of the Frontier in History* and the slide of the economy into depression, Turner quickly became a celebrity among businessmen and politicians.[33] Roosevelt cheered that he had "put into definite shape a good deal of thought which has been floating around rather loosely" and announced that he would incorporate the thesis in the third volume of his own work *The Winning of the West.*[34] Turner argued that after almost three hundred years of European settlement of the New World, progressing from east to west, a specific American character had been formed, characterized by independence, optimism, and a communal spirit—qualities the pioneer needed for survival. "The West, at bottom, is a form of society, rather than an area," Turner wrote. In his opinion, the creation of a democratic and classless America had followed a La-marckian logic: "The history of our political institutions, our democracy, is not a history of imitation, of simple borrowing; it is a history of the evolution and adaptation of organs in response to changed environment, a history of the origin of new political species."[35]

By 1889, Turner argued, the frontier—an abstraction of the remaining land in the West that had not been taken into possession by white settlers—had closed. In a controversial decision, the government had allowed fifty thousand of these settlers to move to a former Indian territory in Oklahoma. A dubious conclusion from the 1890 census allegedly proved that there were no more strips of land available for new settlement. For Turner's audience, the analysis of this situation as well as its remedy seemed evident: The end of expansion, the defining motive of the American nation, had resulted in both an economic and a cultural crisis. There were no domestic markets left to absorb the growing production of American farms and factories. Industrialization and the dramatic social changes that accompanied it threatened to destroy Turner's nostalgic vision of the nation. But expansion could continue abroad, and in the same way American society had adapted to the challenges of the West, it could "expand" again in a moral sense. "That these energies of expansion will no longer operate would be a rash prediction; and the demands for a vigorous foreign policy, for an interoceanic canal, for a revival of our power upon the seas, and for the extension of American influence to outlying islands and adjoining countries, are indications that the movement will continue," Turner believed.[36] It was Theodore Roosevelt who set out to achieve this task and propel the nation into the twentieth century, employing the Panama Canal as his vehicle.

Theodore Roosevelt, Panama, and the "Strenuous Life"

There are few figures in American history as colorful and controversial as Theodore Roosevelt (1858–1919). The legacy of the "politician, soldier, author, hunter, cowboy, and forceful preacher of the balanced banalities," as Robert Wiebe memorably called the former president,[37] has been claimed by his successors both on the Left and the Right, and his personality and rhetoric seem ever present in American public life.[38] Roosevelt (or TR, as he came to be called) shared many of the beliefs put forward by the intellectual expansionists during the 1890s, the decade of his political ascent, and as a quasi-member of the relatively small community of American letters, he knew most of them personally or had at least commented on their works. And yet all attempts to assign Roosevelt's persona and policies to a static ideological framework are bound to fail. Literally, he embodied diverse attitudes—always with a keen eye on expressing them to the public—and became known not only as the epitome of jingoism, the military revitalization of a white, male (and violent) America, but also as one of the first advocates of environmental protection. "I can see him now and hear his unmusical voice saying 'The effort—the effort's worth it,' and see the gesture of his clinched hand and the—how can I describe it? the friendly peering snarl of his face, like a man with the sun in his eyes. He sticks in my mind as that, as a very symbol of the creative will in man, in its limitations, its doubtful adequacy, its valiant persistence amid perplexities and confusions," the British science fiction author H. G. Wells noted after a visit to the White House in 1906.[39] It may not seem too far-fetched to suggest that Roosevelt represented the chances and confusions facing American society at the crossroads of the Victorian Age and the modern era.

Roosevelt was born in New York in 1858. A sickly child, he grew up in an upper-class family of Dutch origins. After graduating from Harvard, he published a historical study of the sea battles fought during the American war with Britain in 1812. Besides his literary work as a historian and author, Roosevelt pursued a political career, first as an assemblyman for the state of New York and later as the civil service and police commissioner of the metropolis. From 1889 to 1896 he published four of the proposed six volumes of his history of the American West.[40] In 1897 TR was named the assistant secretary of the navy and became increasingly involved in foreign affairs. In a letter to Mahan, he emphasized the urgency to annex Hawaii and construct an interoceanic canal: "I believe we should build the Nicaragua canal at once, and in the meantime that we should build a dozen new battleships, half of them on the Pacific Coast."[41] Having resigned from his post in order to fight in the Spanish-American War, the Republican Roosevelt returned to public office as the governor of New York and successfully ran for vice president of the United States in 1900. When President McKinley was killed by an assassin only one year later, the leadership of the nation fell into the hands of the forty-two-year-old Roosevelt. Startled but far from frightened, TR wrote to his friend Henry Cabot Lodge, a fellow expansionist: "It is a dreadful thing to come into the Presidency this way; but it would be a far worse thing to be morbid about it."[42]

As an aspiring politician with an exceptional dose of self-awareness, Roosevelt managed to completely reverse his public image, transforming the quirky, bookish adolescent he was originally perceived as into a resolute and virile leader.[43] When Roosevelt joined the state assembly in Albany at the age of twenty-three, newspaper writers made fun of his high-pitched voice and odd clothing, implying the homosexuality of the young politician. Roosevelt took notice and implemented a comprehensive program of self-improvement including the purchase of a ranch in South Dakota, establishing himself, the former city boy, as a rugged frontiersman and lover of the Wild West, taking up mountain climbing, hunting, and football as his hobbies. Most importantly, he documented these various new endeavors in his books.

At the time, this kind of metamorphosis, although highly unusual for a politician, was not entirely unique. The famous strongmen Eugen Sandow and Bernarr MacFadden encouraged white middle-class men to get in shape, turn their bodies into muscle machines, and display them publicly.[44] Showmen such as William F. Cody, alias Buffalo Bill, used historical events as inspirations and turned them into highly publicized, manly spectacles that then became the public memories of the "real" thing.[45] As the Victorian image of the man as a producer and provider for his family began to fade under the impact of industrialization, interpretations of manhood shifted and adapted to the new social conditions both in the private and the public sphere. While constructed categories such as gender are always contested, historians agree that the 1890s represented a particularly intense phase of such dynamics.[46] The active role of women in the evolving consumer society and their demand for the vote, the assertion and organization of working-class men (including African Americans and immigrants from Southern and Eastern Europe), the rise in the number of single men, and last but not least the changing work patterns, resulting in more white-collar clerical (from clerk) and service jobs for men, all played a part in this process.

In important ways, the remaking of manhood as practiced by Roosevelt or the bodybuilders during the 1890s and the subsequent decades reflected aspects of modern society. It took place within an emerging mass culture, enabling what cultural historian James Gilbert calls "spectatorship masculinity," the emulation of male identities from observing public "heroes" such as the strongmen through photographs in magazines, encountering them in fiction (like Edgar Rice Burrough's Tarzan stories, first published in 1912) or in other, equally constructed public arenas such as Buffalo Bill's Wild West.[47] While the bodybuilders may have evoked ancient (and in that respect anti-modern) ideals—the white male athletic body—they also demonstrated that their bodies could be remade employing a standardized, almost technological process (to build up muscles) that was decidedly modern.[48] Owen Wister's Virginian, protagonist of a bestselling novel published in 1902 and dedicated to the author's former classmate Theodore Roosevelt, transformed the cowboy figure into a modernized hero who sought the American West as a source of strength to confront the perceived ills of contemporary society, joined by an elite of other cowboy-soldiers who took the destiny of the nation into their own hands.[49]

Although Roosevelt's chauvinistic escapades have often been viewed as a reactionary evocation of the manly warrior, they were far more complex. He was not so much concerned with the Victorian image of a man who acted on his own behalf and in accordance with his inner beliefs (and in solitude, if necessary) but instead staged his manhood "before the eyes of a domestic audience."[50] Public relations, in the very sense of the word, were always part of his actions, and the actions themselves were always carefully planned and constructed. In TR's eyes, manliness was not an interior trait but could be imagined, acquired, and acted upon.[51] It was clearly an evolutionary, Lamarckian concept and, for Roosevelt, a metaphor that could be applied to the nation as a whole. "As it is with the individual, so it is with the nation," he wrote.[52] This national audience with whom he was identifying and communicating was the white male middle class, and it would have to undergo the same transformation as Roosevelt himself. America had to become a manly and moral nation.

The future president viewed the apparent problems of industrialized American society in gendered and racialized categories. Effeminacy, one of the key terms in this debate, and decadence were the main obstacles in the country's path to greatness. They were ascribed to women, the "over-civilized" culture of the European aristocracies, and the hedonistic pursuits of the American business class. In his study *The Theory of the Leisure Class* (1899), the social critic Thorstein Veblen subsumed practices characterized by wastefulness and the search for profit under the term "conspicuous consumption." He mentioned high heels, corsets, and other status objects women wore (if only at the request of men)

as examples.[53] Roosevelt feared that a nation "sunk in a scrambling commercialism; heedless of the higher life, the life of aspiration, of toil and
risk" would eventually "go down before other nations which have not
lost the manly and adventurous qualities."[54] Like Adams and Mahan, he
discovered analogies in ancient history. He was impressed by a biography of Alexander the Great written by his friend Benjamin Ide Wheeler,
historian and president of the University of California, and he drew his
own conclusions from it: "Moral corruption ate into the whole social and
domestic fabric, until, a little more than a century after the death of Alexander, the empire which he had left had become a mere glittering shell,
which went down like a house of cards on impact with the Romans; for
the Romans, with all their faults, were then a thoroughly manly race—a
race of strong, virile character."[55] In his massive work *The Winning of the
West,* Roosevelt had presented the Indians as cowardly rapists who were
conquered by the Romans of their time: a more legitimate and manly
"American" race.[56]

In 1898, when the war on Cuba, right in front of the American door-
step, seemed imminent, Roosevelt welcomed the chance to put his ideas
into practice.[57] He resigned from his position as assistant secretary of the
navy and assembled a volunteer cavalry regiment for the battle against
Spain, an unlikely posse consisting of athletic graduates from Ivy League
colleges, Western roughnecks—the type of men who had defeated the
Native Americans in *The Winning of the West*—as well as, surprisingly, a
small number of American Indian, Jewish, Irish, and Italian men who
must have stood out in Roosevelt's mind as exceptions from the dominant traits of their ethnic groups. No African or Asian Americans were
accepted.[58] The Rough Riders, as they came to be called, were to fight
the Spanish "savages" and prove the superiority of the American nation. Roosevelt allowed journalists to observe the battles of his regiment
as "embedded" reporters; he had devised his involvement in the war as
an ample demonstration of "spectatorship masculinity." The war ended
after only three weeks and four battles; the Rough Riders had participated in three of them and were praised by the media as the decisive
factor in the war. As research has shown, three all-black regiments in
cavalry and infantry (commanded by white officers) had played a crucial
role in the military victory, but Roosevelt later toned down his praise for
them considerably and even pointed out their alleged shortcomings in
his account of the war.[59]

The Spanish-American War has often been portrayed as a kind of escapist attempt by policymakers and the media to make their readers and constituency forget the complex challenges facing American society and take part in an unmitigated celebration of jingoism and violence, an anti-modern reaction to the social strains of the closing century, "a triumph of traditional individualism over modern centralizing needs."[60] For Roosevelt, the manly performance of the Rough Riders had been important—after all, it multiplied his prominence—but there was more to his vision of a national revitalization. It required a collective effort ("the power of acting together"[61]), and leaders who fought for the public good—unlike the businessmen who only cared for their own profit. These elite men were the new cowboys from Wister's novel, officers in an army of progress. Similar to Mahan's navy, a modernized army became not only a symbol of Roosevelt's new America but a necessary element in its construction.[62] With this aim in mind, the Army War College was founded in 1901 to train future officers, and Secretary of War Elihu Root introduced plans for a restructuring of the forces in 1901 and 1903.[63] The war with Spain had only been the introduction to a broader program of expansion and reform.

In his famous speech on "The Strenuous Life," Roosevelt summarized his agenda for a future America. Since the closing of the frontier, the country had enlarged its navy, expanded abroad, and was now faced with "the responsibilities that confront us in Hawaii, Cuba, Puerto Rico, and the Philippines," the returning Rough Rider told the all-white, all-male members of the conservative Hamilton Club in Chicago in the spring of 1899.[64] But the task of reconstructing the nation was far from achieved:

> The timid man, the lazy man, the man who distrusts his country, the over-civilized man, who has lost the great fighting, masterful virtues, the ignorant man, and the man of dull mind, whose soul is incapable of feeling the mighty lift that thrills "stern men with empires in their brains"—all these, of course, shrink from seeing us build a navy and an army adequate to our needs; shrink from seeing us do our share of the world's work, by bringing order out of chaos in the great, fair tropic islands from which the valor of our soldiers and sailors has driven the Spanish flag. These are the men who fear the strenuous life, who fear the only national life which is really worth leading. They believe in that cloistered life which saps the hardy virtues in a nation, as it saps them in

the individual; or else they are wedded to that base spirit of gain and greed which recognizes in commercialism the be-all and end-all of national life, instead of realizing that, though an indispensable element, it is, after all, but one of the many elements that go to make up true national greatness.[65]

Roosevelt viewed the work of transforming America as an ongoing, evolutionary process, and there was one master project yet to be achieved, which the soon-to-become president brought up shortly after his assault on the "over-civilized man." It was the interoceanic canal: "We cannot sit huddled within our own borders and avow ourselves merely an assemblage of well-to-do hucksters who care nothing for what happens beyond. Such a policy would defeat even its own end; for as the nations grow to have ever wider and wider interests, and are brought into closer and closer contact, if we are to hold our own in the struggle for naval and commercial supremacy, we must build up our power without our own borders. We must build the isthmian canal, and we must grasp the points of vantage which will enable us to have our say in deciding the destiny of the oceans of the East and the West."[66]

Historians have argued that after the exploits of the late nineteenth century, American foreign policy took a different turn. Under Roosevelt's and Taft's presidencies, the Open Door policy evolved into "Dollar Diplomacy," an approach favoring economic dependencies over colonies that would have to be officially annexed. In 1905, as Emily Rosenberg notes in her elaboration of this argument, TR decided against the annexation of the Dominican Republic and instead installed a "fiscal protectorate" there, guaranteeing American financial control over the island.[67] Kirstin Hoganson finds that anti-imperialist critique was growing after the guerilla war in the Philippines, which had exposed the dangers of the tropics and cast doubt on the goal of constructing manliness through war.[68] Policymakers and the public grew tired of aggressive expansion, she argues: "In their quest to, as Roosevelt put it, 'make man better,' they shifted their energies to the character-building challenges of domestic reform."[69] These analyses are valuable, but they ignore the American endeavor in Panama. While the critical voices mentioned by Hoganson would also be raised with regard to the construction of the Canal, Roosevelt managed to silence these opponents and impose his vision on the project, as he had explained it in "The Strenuous Life." The Isthmus was far more than a foreign territory suitable for economic utilization; it became the digging ground for the expansionists' vision of a new America.

In 1903, however, the Canal project was deadlocked. Colombia demanded more money in return for the ratification of the Hay–Herrán Treaty, and Roosevelt was not willing to give in. The circumstances that eventually brought about the existence of the Panamanian nation and made possible the century-old dream of an interoceanic waterway form one of the most unusual and controversial chapters in the history of American foreign policy.[70]

Among the Panamanian elite, the Colombian regime was not well regarded. The conservative rulers in Bogotá, devoted to fostering Spanish traditions, had cared little for their northern province and looked down on the multiethnic country. From the perspective of the Isthmus, the Colombian capital high up in the Andes was far away, and the more liberal-minded Panamanians felt much closer to the Caribbean and the United States. With Colombia in disarray after a civil war and the building of a Canal in sight, an independence movement with strong historical roots began to reassert itself once again. Lobbying for this group in the United States was none other than Philippe Bunau-Varilla, the Frenchman whose sole objective was to see the Canal built. Indirectly, he suggested to the American government that a revolution in Panama would succeed if backed or at least tolerated by the new superpower. Roosevelt liked the idea; an end to the standstill seemed in sight. Though he was generally averse to revolutions, the president hoped that an independent Panama would guarantee stability in the region—and grant the Canal to the United States. Later, Roosevelt would repeat over and over again that the small Latin American country had seen fifty-three failed revolutions in the fifty-seven years since 1846.[71] "Much as we wanted Panama, the Panamanians wanted us more, and if there was one thing experience had taught them it was how to organize a revolution," one of the Panama authors wrote.[72] There was no official affirmation, but Bunau-Varilla was given hints that the United States would see to it that Colombia did not strike down the revolt.

On October 30, 1903, the battleship USS *Nashville* left Jamaica for Colon on a vaguely defined errand. Four days later, the Panamanian junta, led by Manuel Amador Guerrero, a physician affiliated with the Panama Railroad, declared the country's independence from Colombia. "There will be some lively times in carrying out this policy," Roosevelt wrote to his son Kermit.[73] A battalion of Colombian soldiers had already landed in Colon, in spite of the presence of the U.S. Navy. But unable to reach Panama City, the sight of the uprising, by railroad—it had deliberately been put out of use—they remained inert, and the Panamanian revolution succeeded with only a single combat death. Additional American gunboats arrived on the scene to prevent any Colombian moves. With a considerable amount of luck, Roosevelt's goal had been achieved.

Bunau-Varilla then introduced himself in Washington, D.C., as the official emissary of the Panamanian state, ready to negotiate a new Canal treaty based on a draft by Secretary of State Hay. Afraid that the Senate would not approve the treaty, Bunau-Varilla, who had designed the new Panamanian flag himself, single-handedly applied some changes to Hay's draft that were favorable to the U.S. The Hay–Bunau-Varilla Treaty enlarged the original six-mile-wide Canal Zone to ten miles, yielding an area of five hundred square miles, and extended the 100-year lease of the land from Panama to perpetuity. The new country was placed under protection of the United States, which gained the same "rights, power and authority" in the zone it would have had "if it were the sovereign of the territory"; formal sovereignty was assigned to Panama.[74] It was a wording that granted the United States "a virtual protectorate over the new country" and would strain the relationships between the two nations for decades to come.[75] The United States was also given far-reaching administrative control of the towns of Colon and Panama City for the enforcement of sanitary measures and the "maintenance of public order."[76] Panama had become a "potential colony of the United States."[77] When junta leader Amador finally arrived in Washington by train on November 18, he read the treaty, signed three hours earlier, for the first time. An amused Bunau-Varilla noted in his memoirs: "Amador was positively overcome by the ordeal. He nearly swooned on the platform of the station."[78]

Despite the devastating restrictions, the Panamanian side was in no position to turn down the treaty and thus accepted it two weeks later. Amador was officially elected president. The United States, now backed by a new legal framework, had a dominating influence on the fate of the country, after having intervened in its affairs ever since the building of the railroad. Prior to 1920 there was little successful private American investment in Panama except for the large banana plants set up by the United Fruit Company.[79] Politically, the United States supported Amador's Conservative Party, which catered to a white elite, and in late 1904 shattered the aspirations of his potential opponent, an army general and supporter of the popular Liberal Party—the political basis for

Panamanians of diverse ethnic origins—by dismantling the small army and replacing it with a police force.[80] In 1906 American election "supervision" helped to keep the Conservatives in power.

In the United States, the Democrats in the Senate, led by John Tyler Morgan, and parts of the media expressed strong criticism of Roosevelt's procedure. In his message to Congress on January 4, 1904, the president placed the events in Panama, which had not been part of any official plan, within his framework of an aggressive foreign policy. "If ever a Government could be said to have received a mandate from civilization to effect an object the accomplishment of which was demanded in the interest of mankind, the United States holds that position with regard to the interoceanic canal," Roosevelt declared.[81] In this message and the one given a year later, the president shaped the Monroe Doctrine of 1823 into a more modern and assertive version referred to as the Roosevelt Corollary to the Monroe Doctrine, laying the foundation for the U.S. role as the "policeman of the world." Within two decades, the "big brother" would exert economic or military control over fourteen of the twenty countries in Latin America.[82] "We, in effect, policed the Isthmus in the interest of its inhabitants and of our own national needs, and for the good of the entire civilized world."[83] Once again, Roosevelt had relied on a moral argument. The Senate, unwilling to see the Canal project fail, approved the Hay–Bunau-Varilla Treaty by a majority of 66 to 14 votes. In the fall, Roosevelt secured a second term in office.

Public chiding of the U.S. role in the creation of Panama resumed in 1911, a few years prior to the opening of the Canal, when Roosevelt made a controversial comment regarding the events in a Charter Day speech at the University of California at Berkeley. The press printed different versions of Roosevelt's remark, and the exact wording remains unclear. Joseph Bucklin Bishop, secretary of the Canal Commission and subsequently Roosevelt's biographer, quoted the former president as follows: "If I had followed traditional, conservative methods I should have submitted a dignified state paper of probably two hundred pages to Congress, and the debate on it would be going on yet; but I took the Canal Zone and let Congress debate and while the debate goes on the canal does too."[84] The offensive phrase "I took the Canal Zone" prompted criticism both from the domestic opposition and from Colombia. In 1914 the Wilson administration offered the Latin American country an apology and a payment of $25 million as a compensation for the Canal benefits it would miss. Roosevelt was furious, and in the end, the Thomson–Urrutia Treaty was ratified only in 1921, two years after his death and without the apology.[85] When they explained the Canal to the U.S. public, the Panama authors, most of them writing after Roosevelt's speech, took diverse views on this political issue, but in general they did not devote a lot of attention to it. Logan Marshall stated with regard to the Panamanian revolution that the United States had been "entirely free from participation in it,"[86] while fellow writer Willis John Abbot admitted that "about the international morality of the proceedings which created the relations now existing between the United States and Panama perhaps the least said the better."[87]

The controversy surrounding Roosevelt's actions and their justification is significant not only because it would resurface in the debate of the 1970s but also because it demonstrates how important the Panama Canal was for him. Months before the speech, he had written to his successor, William Howard Taft: "I have always felt that the one thing for which I deserve most credit in my entire administration was my action in seizing the psychological moment to get complete control of Panama. Incidentally, it was one of the things for which I was most attacked."[88] Speaking in front of an audience of historians at the Panama-Pacific International Exposition in San Francisco in 1915, he told his listeners that there was "not one action of the American Government in connection with foreign affairs, from the day the Constitution was adapted down to the present time, so important as the action taken by the Government in connection with the acquisition and building of the Panama Canal."[89] The Canal building was the key in his quest for "true national greatness." And in 1904, after the creation of Panama, Roosevelt's involvement with the project had only just begun.

Making the Dirt Fly: Critics and Crises on the Isthmus

The American public had long been waiting for this moment: The building of the Panama Canal was about to begin. On April 8, 1904, President Roosevelt appointed a new Isthmian Canal Commission (ICC), headed by Admiral John G. Walker.[90] A month later, the French possessions, the Panama Railroad, and the Canal Zone were transferred to their new owner, the U.S. government. The territory was ruled by executive order; it was not part of the United States, and federal laws did not apply to it. The commission reported to Secretary of War Taft. The railroad official John Findley Wallace was named chief engineer. He and his successor John F. Stevens were exposed to enormous pressure "to make the dirt fly."[91] After years of back and forth, the press as well as the politicians demanded quick results. There was little patience for the lengthy preparations necessary to get the project started.

The seven-member Canal Commission was ill-equipped to satisfy these demands. Working out of Washington, D.C., it had no real control of the work done in Panama. Vice versa, Wallace, who was not part of the commission, had to get even the smallest order for material approved by the bureaucrats back in the United States. This process took weeks, and soon enough, complaints of red tape abounded. "There was confusion, procrastination, lack of system everywhere," the secretary of the third commission, Joseph Bucklin Bishop, wrote in hindsight.[92] His son Farnham, another Panama author, claimed to know the reason for the malaise: "An army commanded by a commission of seven men has exactly six generals too many."[93] Apart from the pitfalls of hierarchy, no decision had yet been reached whether the waterway was supposed to be built as a sea-level canal—as in the original French plan—or as a canal with locks.

For the first time, Americans confronted Panama not only as a transit stop or reference to diplomatic entanglements, but as an actual place in the tropics. This encounter with an unknown, often threatening environment created another negative impression, which can also be traced in the writings of the storytellers. "The streets, unspeakably dirty and mud-filled, swarmed with naked children; the ugly frame houses rested on piles, under which greenish slimy water formed lagoons. Such dilapidation and desolation!" wrote Marie Gorgas, wife of William Crawford Gorgas, the chief sanitary officer of the American force, on a visit to Colon.[94] Observing the street life in Panama City, another shocked visitor reported: "In all the world there is not another city so depressing as Panama. Natives and aliens go through its streets with dragging feet and saddened faces. I saw a cab-horse, standing in front of the dismal public square, move a few inches forward so that he might lean against a telegraph pole—and the action was so much in the spirit of the town that it did not jar upon my perceptions as remarkable until long afterward. We wondered how American white men of decent tastes could stay more than six weeks in such a hole and keep their reason."[95] Comments such as these on life in the Panamanian cities, where many people of diverse ethnic origins lived, were linked to a pervading racism. Its basis could be found in the expansionists' propagation of the supremacy of American civilization, supported by the demeaning displays of foreign people (and African Americans) at world's fairs and in magazines and stereographs. The most blatant manifestation of these racist attitudes was the segregation of the labor force.

For the construction of the Canal, large numbers of unskilled laborers were needed. They were mostly recruited from the West Indies, especially Jamaica and Barbados. Around 5,000 Caribbean workers had come to Panama during the construction of the railroad, and another 50,000 were employed by the French. As Michael Conniff states, about 150,000 West Indian laborers toiled on the American Canal project during the entire construction decade.[96] They worked in excavation as well as in the service sector of the Canal Zone. The overwhelming majority of skilled workers such as engineers and machinists were white Americans; there were few Native Panamanians on the force. As a consequence, an odd differential was created in the Canal Zone, the dichotomy of *gold* and *silver roll*. It was named after the material payment: gold employees received U.S. dollars, while silver workers were paid in Panamanian dollars worth less than half—in addition to their lower payment in absolute figures. Although this system officially connoted a distinction between skilled and unskilled work, it was in fact a racial determinant. Black Americans worked only in small numbers on the Isthmus — allegedly "because of the objections that would be raised to their removal in large numbers from the farms"[97]—and white Americans were not recruited for unskilled labor. From late 1906 on, when all colored workers were put on the silver roll, the difference between white, "civilized" Americans and black, "uncivilized" West Indians was therefore fairly clear-cut—except for a large number of laborers from Southern Europe, notably Spain, Italy, and Greece, who were recruited "to act as the corps d'elite of the Silver force."[98] Julie Greene has shown that these workers were sometimes regarded as white, sometime as nonwhite.[99] Their fluctuating racial identity reflected the status of immigrants from Eastern and Southern Europe in the United States.

Silver employees were not entitled to benefits such as free housing, paid vacation, and social entertainment. As a further discrimination, a full-fledged Jim Crow system of segregation was enforced in the Canal Zone. "Panama is below the Mason and Dixon line," one author commented.[100] When a colored American citizen who was actually paid in gold—the whole case being a rare exception—tried to cash a money order at a teller reserved for gold employees, he was rebuked. "As you are employed on the gold roll, of course you have a technical right to transact business on the gold roll side of the post office, although, as you are of the negro race, it would seem that you would avoid disputes of this character and save yourself and others annoyance if you would transact business on the side where others of your race transact their business," the chief engineer's executive secretary responded to his complaint.[101]

The negative comments on the Canal Zone mentioned earlier were not only rooted in a racist perception of the foreign country but on actual threats as well. Outbreaks of yellow fever had severely harmed the French project, and malaria was an enduring problem on the Isthmus. In the mind of the American poet James Stanley Gilbert, "Panama's most famous bard and cruel critic,"[102] the perceived and real dangers of the jungle all mingled together. With an almost sadistic sense of discomfiture, Gilbert captured the terror of the tropics:

> Tis the land where all the insects breed
> That live by bite and sting;
> Where the birds are quite winged rainbows bright,
> Tho' seldom one doth sing!
> Here radiant flowers and orchids thrive
> And bloom perennially—
> All beauteous, yes—but odorless!
> In the Land of the Cocoanut-Tree.[103]

Roosevelt had urged commission chairman Walker to employ "the very best medical man in the country" for the fight against disease.[104] When Gorgas, who had done pioneer work eradicating yellow fever in Cuba's capital Havana, was sent to the Isthmus, he encountered the same obstacles as Chief Engineer Wallace. His requests went unanswered or were denied by the commission, whose members expressed little sympathy for his plan to get rid of mosquitoes in the Canal Zone. Despite some successful measures, an outbreak of yellow fever lasting from November 1904 through September 1905 could not be prevented. The disease that Gilbert had described with the help of all his morbid talent ("Your mouth will taste of untold things, / With claws and horns and fins and wings; / Your head will weigh a ton or more, / And forty gales within it roar!"[105]) killed dozens of workers, and hundreds of others were so scared that they quit their job and returned to the States.[106] The crisis on the Isthmus had reached its peak.

In April 1905 Roosevelt had already appointed a new Canal Commission headed by Theodore P. Shonts, a former railroad manager. The
number of commission members was not changed, but Wallace became
one of them, and together with the new Canal Zone governor Charles
Magoon, a more effective "triumvirate" was now in charge.[107] In June
Wallace resigned to take an industry job, and the construction engineer
John Stevens assumed his position. With Roosevelt's support, he assigned greater priority to the sanitation work and within his two-year
term managed to erect almost all of the infrastructure necessary to
begin the actual construction of the Canal. The old railroad dating back
to 1855, which seemed "nearly at a standstill," was replaced by a new
double-track version.[108] Shonts rallied support for the project, telling his
expert audiences: "We are building the 'Roosevelt Canal.'"[109] Walter
Pepperman, who had been the administrative chief of the second
Canal Commission, published his own revisionist account of the Canal
construction in 1915. It differed significantly from the works of his fellow
Panama authors. Whereas the majority of the storytellers extolled the
achievements of the third commission under Goethals, Pepperman
tried to give most of the credit to Shonts and Stevens, "the railroad
men," who in his opinion had laid the foundations for success. His book
demonstrates that individual authors pursued their own agendas,
thereby contesting (or confirming) the dominant Canal discourse.

Despite the visible progress, domestic criticism of the Canal project
continued and culminated in an article appearing in the *Independent*
magazine on January 4, 1906, entitled "Our Mismanagement at Panama."[110] The author, Poultney Bigelow, was the son of a former U.S.
ambassador to France, John Bigelow. He pretended to take a close look
at the Canal project and was inspired by the style of investigative journalists whom Roosevelt would later dub the "muckrakers." Exposing
the ills of industrialized society and an unchecked capitalism, these writers hoped to support political reform movements through their work.
Boosting the circulation of new magazines, they were highly useful
commercially to their editors.[111] While Bigelow's methods and motives
must be put into question—his opponents calculated that he had spent
only twenty-eight hours and ten minutes on the Isthmus[112]—he was
clearly trying to imitate the muckrakers, reporting on red tape and political favoritism, the complaints and unfair treatment of the West Indian
laborers, and the insufficient sanitary conditions in Panama's cities. He
claimed that official reports such as the one published after Secretary of

War Taft's visit to the Canal Zone ignored the problems. Finally, he touched on a rather delicate issue: "On the occasion of my visit the clergy of the Isthmus were loud in protest because the United States authorities had imported at considerable expense several hundreds of colored ladies. Prostitutes are not needed on the Isthmus—and if they were there is no call to send for them at the expense of the taxpayer."[113]

In Washington, Bigelow's report raised a storm of protest. Roosevelt asked Taft for a report that refuted the journalist's allegations and that was also published in the *Independent*. Taft aimed at Bigelow's doubtful procedure—his short visit and use of biased sources, such as the American resident Tracy Robinson, a notorious misanthrope who would later be utilized by the Panama authors as a witness of moral corruption during the French era.[114] The magazine printed another, more positive report from Panama, this time bylined by one of their own reporters (Bigelow worked as a freelancer).[115] The journalist received a chance by the mass magazine *Cosmopolitan* to repeat his criticism. The editors took care to note, though, that his previous piece had relied "on a very superficial examination."[116] Praising the Canal work, the Panama authors later castigated Bigelow's pieces as examples of "the bilious effusions of yellow journalism and the mendacious maunderings of sensation mongers" threatening to kill the project.[117] The journalist himself, Pepperman exclaimed, was one of the "human mosquitoes" on the Isthmus.[118]

In 1906 Bigelow's criticism was more or less representative of the pervading sense of crisis regarding the Canal project. It also prompted an inquiry by the Senate's Interoceanic Canals Committee, headed by the fierce proponent of a canal through Nicaragua, John Tyler Morgan. This feeling of crisis, as we have seen, was based on actual problems of administration and disease management but also on the underlying resentment and fear of the alien country of Panama and its multiracial society. It is also plausible that the possibility of failure—signs of the French disaster were visible all over the Canal Zone—and misgivings about the greatness of American engineering and planning played a role. The rapid technological changes of the age were disturbing, and a project of this size and on a foreign territory was almost unimaginable. The Panama authors described the crisis mostly in hindsight, and it was a useful first act in their plotting of an American success story on the Isthmus. But in the early days of construction, Roosevelt's vision of the Canal as the restorer of a manly, national spirit was nowhere in sight. It was up to the president to bring it back.

"A Monster Electric Battery": TR Saves the Canal Project

To quiet the critics and oversee the work that had been done, President Roosevelt decided to take a trip to the Canal Zone. He did not wait until the end of the rainy season, so none of the Canal's foes could object that he had seen only the sunny side of the Isthmus. It was the first time in American history that an incumbent president left the country while in office, and it would remain Roosevelt's only visit to Panama—he never saw the completed Canal.[119] As a passenger on the USS *Louisiana,* the navy's largest battleship, Roosevelt embarked on his two-week trip on November 9, 1906, accompanied by his wife Edith Carow (nickname "Mother"). Five days later, almost a day ahead of schedule, they arrived in Panama. When Roosevelt landed, the *New York Times* reported, there was no official entourage to greet him, and the president introduced himself to the nearest person as "Mr. Roosevelt of Washington."[120]

Roosevelt carefully staged his visit to the Canal Zone, presenting to his audience at home the image of an adventurer and investigator who despite his role as a dignified statesman constantly ignored the official protocol in order to meet the Canal workers, register their complaints, and remind them of their share "in adding renown to the nation under whose flag this canal is being built."[121] He disregarded a festive luncheon and instead sat down with the American gold-roll men for a 30-cent meal in one of the mess halls. En route to Culebra Cut, the site of the most difficult excavations, the train was stopped and he climbed up to the "cockpit" of a ninety-five-ton Bucyrus steam shovel. The photograph of Roosevelt sitting at the lever of the gigantic dark machine in a white suit, a powerful leader in the literal sense of the word, became the emblem of his trip and is discussed in greater detail in a following chapter.

While the press praised his spontaneity, there are clear indications that instead of actually being spontaneous, Roosevelt had worked hard to convey this impression. He made sure that pictures of his various inspections on the Isthmus were being taken not only by the news media but also for the official record. His message to Congress became the first illustrated delivery of its kind. The inclusion of photos in the printed version had the purpose of not only reinforcing his arguments, but also of presenting him as an innovator. When Roosevelt gave a speech to the laborers—which would also be attached to his official message package—he received questions from the audience. Stenographer Mary Chatfield, who was present at the ceremony, later conjectured in her memoirs: "His voice gave out several times during his speech. He was often interrupted by a very tall man who, I presume, was hired to do so."[122] Just as Roosevelt had constructed his own persona, he ventured to Panama to save the project from failure and to inscribe into the Canal the meanings he had originally conceived for it. His visit is best viewed as a public relations effort. "I go back a better American, a prouder American, because of what I have seen the pick of American manhood doing here on the Isthmus," Roosevelt told the laborers.[123]

In his message to Congress, the president gave detailed reports on the preliminary work accomplished on the Isthmus, including Canal construction, railroad management, sanitation, health care, housing, food, and recreation facilities, as well as the improvements in the infrastructure of Panamanian cities. He addressed grievances, even by the West Indian laborers, joked that "but a single mosquito, and this not of the dangerous species,"[124] had crossed his path during the trip, but found harsh words for "doubting Thomases" and other critics of the enterprise: "There remains an immense amount of as reckless slander as has ever been published. Where the slanderers are of foreign origin I have no concern with them. Where they are Americans, I feel for them the heartiest contempt and indignation; because, in a spirit of wanton dishonesty and malice, they are trying to interfere with, and hamper the execution of, the greatest work of the kind ever attempted, and are seeking to bring to naught the efforts of their countrymen to put to the credit of America one of the giant feats of the ages."[125]

Roosevelt's trip was not only seen as a success by news media and magazines, but it also became the turning point of the project in the writings of the Panama authors. "A man had come to Panama, and mingled among the men of the Isthmus for a span of hours, and it was as though a monster electric battery had galvanized the life of the jungle," the author Hugh Weir wrote, employing the same kind of technological vocabulary the Progressive intellectual Herbert Croly used in his description of Roosevelt as a "sixty-horse-power moral motor-car."[126]

At the time of the president's visit, important decisions regarding the future Canal work had already been made. Even though an independent study group had favored a sea-level waterway, the Canal Commission, with the support of the government, finally decided on a cheaper and less time-consuming version with locks. In June 1906 Congress approved the plan. During his trip to the Isthmus, Roosevelt signed an executive order that once again redistributed the powers of the commission.[127]

The chairman was to be given more authority, and the chief engineer would be entitled to act on the chairman's behalf in the Canal Zone. The office of the governor of the Canal Zone was abolished. When Chairman Shonts resigned in January 1907 to take an industry position in New York, it looked as if Chief Engineer Stevens would assume his position as well. But only a few days later, Stevens complained about his job in a letter to Roosevelt, provoking his dismissal. After several disappointments concerning managers with a background in private business, the president picked a former member of the Army Corps of Engineers and a West Point graduate, Lieutenant Colonel George Washington Goethals, as Stevens's successor. He was named both chief engineer and chairman of the third Canal Commission on April 2, 1907, and therefore held far greater powers than any other previous manager on the project, earning him the characterization of a "benevolent despot."[128] Along with Goethals, two other army engineers were appointed commission members and heads of engineering divisions.

Roosevelt had openly favored the building of the Canal by private contract during his trip to Panama. Bids were already starting to come in, but the new "army regime," as it would often be called, suggested that the employment of laborers and the Canal construction would be best carried out under the auspices of the government, as had been the practice so far. In addition, the drafting of the contracts turned out to be a controversial political issue. "Under these circumstances, and also in view of the contemplated change in the administration of canal affairs, as well as his desire not to hamper a new administration with a contract it did not view with favor, the President concluded that he would reject all bids," Goethals explained later.[129] Something extraordinary had happened: TR had put the project back on track by means of his cleverly orchestrated performance on the Isthmus. Through a series of unpredictable, more or less intended moves, the building of the Canal would then proceed as a genuine government endeavor, led by public managers who represented the kind of manly "national life" to which Roosevelt had referred in his speeches. With the preparatory work already done, the Canal construction advanced rapidly under Goethals. As is shown in the following chapters, the Panama authors based their interpretations on the Canal image and administration created in 1906 and 1907, and as Roosevelt had envisioned it, they viewed the success story in Panama as a model for American society as a whole. The crisis on the Isthmus had been overcome, and so could the social challenges facing the nation.

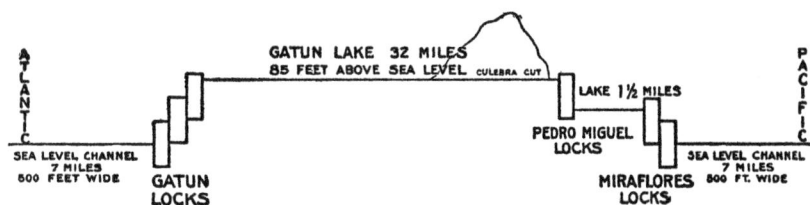

Figure 1. The different levels of the Panama Canal (reprinted from William Rufus Scott, *The Americans in Panama* [New York: Statler, 1912], 156).

The Canal was built by the government, but not by the army itself. The workers as well as the management were employed by the Canal Commission and the Panama Railroad (P.R.R.) Company, which remained a privately run enterprise but was now owned by the government. This setup allowed the P.R.R. to make profits and enjoy financial flexibility. Besides owning two hotels and a steamship line, it was the parent of the so-called commissary, a chain of stores and other enterprises set up to take care of the laborers' needs. The commissary became the primary economic power in the Canal Zone and all of Panama, with its sales of $4.5 million (in 1908) easily outnumbering the revenues of the entire Panamanian state (less than $2.5 million).[130] The reason for this development was rooted in the tariff deals the United States had struck with Panama. Originally, the Panamanian cities were designated to become free ports, while imports into the Canal Zone from other than the United States would be subject to the highly protectionist Dingley Tariff of 1897. This arrangement, depriving Panama of any income from customs, was changed in favor of the Taft Agreement, named after the secretary of war and valid throughout the construction period. Panama was allowed to collect tariffs on all but "necessary and convenient" imports, at an average rate of 10 percent. Between Panama and the Canal Zone, trade was free. The Dingley Tariff for the zone was abolished.[131] The agreement proved devastating for the Panamanian merchants, who could not compete with the prices of the commissary since luxury goods were imported duty-free directly into the zone. To make things worse for the merchants, the commissary was not necessarily dependent on profits and therefore kept prices low.[132]

The Canal itself would feature six sets of locks, three on the Atlantic side (Gatun) and one (Pedro Miguel) plus two (Miraflores) on the Pacific. The ships passing through the seaway were elevated to a height of eighty-five feet above sea level (fig. 1). Next to the Gatun locks, a giant

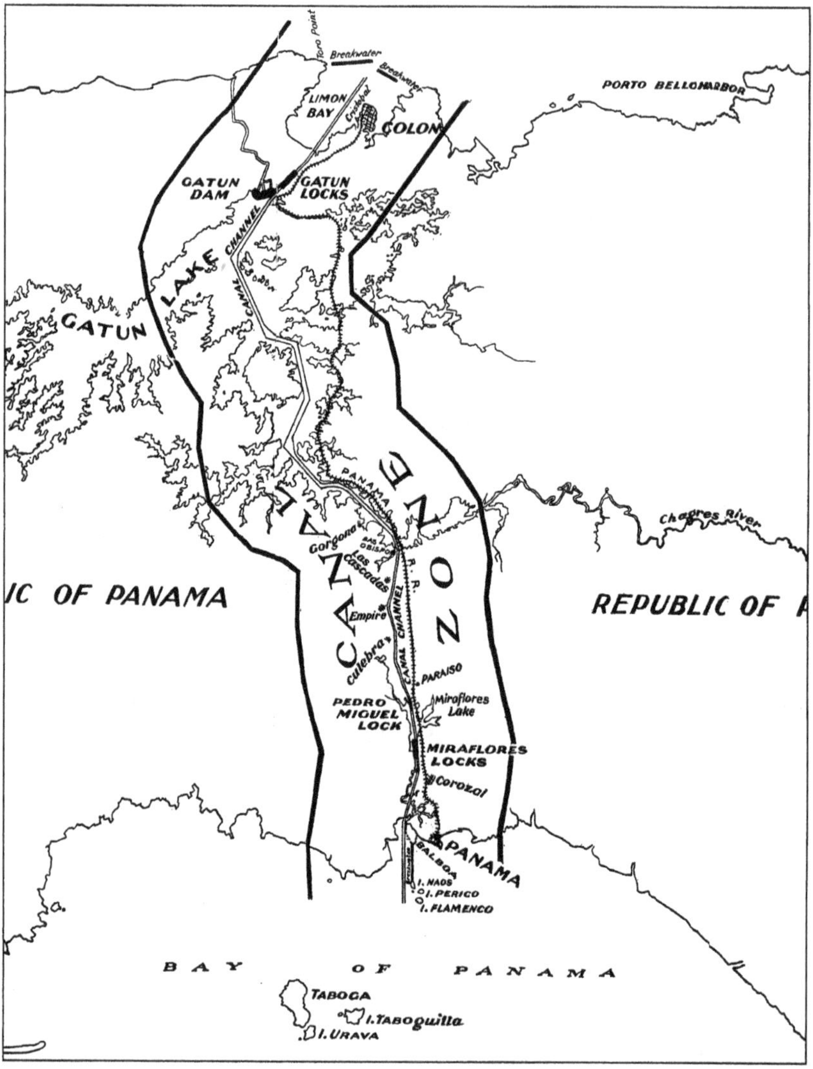

Figure 2. Historical map of the Panama Canal and the Canal Zone (reprinted from Scott, *The Americans in Panama*).

dam would prevent the Chagres River from flowing into the sea, thereby creating an artificial lake on former jungle land. Between the spot where the Chagres flowed down from its source and the Pedro Miguel lock was the continental divide. This was the nine-mile stretch where most of the actual digging had to be done—the infamous Culebra Cut (fig. 2). The operation of the Canal would require little power since the water from Gatun Lake would flow down the locks and into the oceans by the force of gravity and in turn lift up the ships in the lock chambers.[133] There was enough water in the reservoir to "feed" the locks, as the upper Chagres originated in mountainous territory. When the Canal was completed, the transit took eight to ten hours.

Summing Up: The Prelude to a Success Story

U.S. intervention in Panama, triggered by the pursuit of an interoceanic canal and legitimated by the Monroe Doctrine, dates back to the construction of the railroad in the 1850s. After the failure of the French project, the plea for a canal became part of the larger agenda of American expansionism. Economic motives had always played a role in the conception of the waterway as a trade route between the oceans. The canal fit well into the scheme of gaining access to foreign markets—the apparent cure for the social and economic crises of the 1890s. But intellectuals like Mahan, Adams, Turner, and Roosevelt had other objectives in mind as well. Analyzing the history of Western society, they became advocates of expansion for strategic as well as moral reasons. An enlarged navy would elevate the United States to the level of a military superpower. Instead of bellicose opposition, which may have reminded the expansionists of the social upheaval caused by aggressive business competition at home, the Western nations would pursue a global peace order of powerful networks—at the cost of other peoples and races. The Canal was destined to become the great enabler of these ideas.

For the thinkers, expansion was the defining attribute of American society, and after the alleged closing of the frontier, the building of the Panama Canal offered the chance to avoid stasis and renew this national impulse of historical, technological, and moral dimensions. Evolutionary theory promised the continuing ascent of the Anglo-Saxon race, if only a common will could be mustered to hold back the devastating effects of "over-civilization" at home, represented by a hedonistic elite seeking individual profits. In his effort to restore the "manliness" of the nation, Roosevelt transformed the pioneer values of domestic expansion into a modern version of public idealism—the cowboy-turned-administrator, leader of the ever-expanding social empire of the future. By his own example, Roosevelt embodied this Lamarckian logic, demonstrating to his middle-class audiences that a new man and a new nation could literally be constructed.

Assisted by the incidental events leading to the creation of Panama, Roosevelt utilized the Isthmus as a stage to enact this vision of personal and national transformation. The construction of the Canal commenced, but for the first three years it looked like a failure. TR's visit to the Canal Zone symbolized his triumph over the critics of the project—now identical with "the men who fear the strenuous life" he had chided a few years earlier—even though it had little relevance for the logistical success of the Canal building. The subsequent installation of the "army regime" under Chief Engineer Goethals was a logical (if largely unintended) step toward the collective spirit and management the expansionists had called for. The Canal remained a genuinely public project—the greatest of all times—forming the basis for the Panama authors' endorsement of an expanded role of the government and a society controlled and led by experts. The discriminatory constitution of the Canal Zone state, on the other hand, intensified Panama's economic and political dependency on the United States, foreshadowing a century of bitter conflicts.

2

American Triumph

Explaining the Canal Project

When most of the Panama authors were writing their books and articles, the Canal had almost become a reality. But could the average American actually comprehend what had been achieved in the Isthmian jungle, thousands of miles away, employing technological and logistical innovations on a scale never seen before? For this task, the storytellers had to construct the Panama Canal all over again through their writings and images. They had to explain to the readers in the United States why it was relevant. Their writings, along with other travel accounts, photographs, and exhibitions at world's fairs depicting the new and often exotic dependencies of the United States, constituted what the historian Ricardo Salvatore has called the "soft machinery of empire."[1] President Roosevelt had envisioned the Canal as a national endeavor, charged with meanings far beyond the commercial benefits that an interoceanic waterway had promised for centuries. With Roosevelt's and the other expansionists' ideas in mind and well aware of the challenges facing the American middle class—their readership—at the beginning of the twentieth century, the Panama authors embarked on their mission. As a result, they interpreted the Canal project as a defining moment in the history of the nation, and as a utopian model for its future.

From Pest Hole to Health Resort:
The Containment of Disease

In the early years of the Canal construction, tropical diseases were perhaps the most threatening aspect of life and work on the Isthmus and
powerful enough to let the project fail. The resident poet James Stanley
Gilbert expressed no doubts regarding the unhealthy nature of the
Panamanian jungle. In one of his most frequently quoted works, he
exclaimed:

> Beyond the Chagres River
> Are paths that lead to death—
> To the fever's deadly breezes,
> To malaria's poisonous breath!
> Beyond the tropic foliage,
> Where the alligator waits,
> Are the mansions of the Devil—
> His original estates![2]

Only a few years later, historian Charles Francis Adams, brother of
Brooks and Henry Adams, noted that the Canal Zone was "to all appearance an agreeable winter health-resort."[3] How was this radical
change in perception possible? Adams continued: "Thus the Canal
Zone is an object lesson, and the Canal itself a monument; for the last
was, humanly speaking, made possible by a medical triumph, the like of
which in importance to mankind has not been equalled since the discoveries of anæsthetics and antiseptics."[4] The defeat of yellow fever and, to
a lesser extent, malaria became the opening chapter in the success story
written by the Panama authors.

Throughout the nineteenth century, medical science made important advances, and yet the origin of many diseases remained a mystery.
The most significant drop in death rates in the British military at home
and around the world, historian Philip Curtin argues, occurred during
the midcentury decades and was achieved by empirical measures (such
as preventive medicine and improved water supply) unrelated to specific
medical research.[5] During the American Civil War, the centralized lo-
gistics of sanitation, applying everywhere the same kind of general measures, had proven successful. Faced with recurring health crises such as
the yellow fever epidemic in Memphis in 1873 and in the Mississippi
Valley in 1878, more and more cities decided to establish municipal
health boards authorized to carry out vigorous measures.[6] It was widely

believed that most infectious diseases were transmitted by filth, touch, and poisonous gases (miasmas, or bad air). Politicians were convinced that "massive cleanup campaigns" in the industrialized cities were the best response.[7] From the 1880s on, the revolutionary but slow success of the germ theory, stating that infections such as cholera and typhoid fever were caused by bacteria, demonstrated the need for sewage systems and the regulation of food suppliers. Beyond disease containment, health propaganda also became part of the middle-class leaders' attempt to control the life of the urban lower classes. The instructions regarding hygiene regulations given to immigrants from Europe upon arrival in the United States may serve as an example.[8]

"Miasmatism" continued to serve as an explanation for the spread of yellow fever and malaria, even though scientists had expressed doubts about the theory for many years. Mosquitoes had long been named as the possible transmitters: in 1848 by a physician from Alabama and again in 1882 by a professor of obstetrics in Washington, D.C.—the same doctor who had been present at Lincoln's assassination.[9] Historian Richard Evans, in his analysis of the cholera epidemic in Hamburg in 1892, shows that the acceptance of new medical theories also depended on social and political processes, not on scientific proof alone. As the construction of the Panama Canal commenced, the *Scientific American*—of all publications—still referred to "unhealthy gases and poisonous vapors" on the Isthmus as the sources of disease: "In the early days of the canal history, the white mist that rose from the disturbed soil of the isthmus was far more disastrous in its killing effects than the mist of the ocean. It rose from the soil like incense from a brazier. It carried with it from its underground prison all the poison of putrefaction and wherever it inclosed its victims, there fever and death followed."[10] The term "malaria" literally meant "bad air," based on the miasma theory.

The French era in Panama brought back bad memories. On a hill near Ancon "stood for years a yellow fever memorial,"[11] the residence of the first director-general of the Canal project, named *La Folie Dingler*. His wife and two children had died of yellow fever. Ignorance of how the disease was spread had even transformed the French hospitals into dangerous places: In order to prevent ants from climbing up the beds, the legs were immersed into water-filled dishes—an ideal habitat for the mosquito larvae to breed in. Only inches away from their birthplace they had a good chance of finding a patient infected with the disease and then spreading it to other humans.[12]

William C. Gorgas, chief sanitary officer of the American force, arrived on the Isthmus in 1904.[13] He was an army physician experienced
in the battle against tropical diseases. While working in Fort Brown on
the Rio Grande in 1882, he and his wife had been infected with yellow
fever. Both survived and became therefore immune. At that time, Marie
Gorgas reports in her memoir, even bananas and oranges were sus-
pected of transmitting the disease.[14] Gorgas himself was not a strong
supporter of the mosquito theory until he was ordered to Cuba during
the Spanish-American War. In the capital of Havana, which was later
put under American administration, he applied the usual methods
known from the United States against yellow fever—without any success. "All our cleaning up and expenditure not only had not bettered
things, but had even made them worse," he wrote later.[15]

By the order of George Sternberg, a leading bacteriologist and the
surgeon general of the U.S. Army, a team of four researchers came to
Havana in 1900, headed by Walter Reed. Their mission was to test the
mosquito theory. Reed followed the lead of the Cuban physician Carlos
Finlay, who assumed that the *Stegomyia* mosquito was responsible for the
spread of yellow fever.[16] He and his team worked with Spanish volunteers who would let themselves get stung for money. Later, Americans
would do the same, allegedly for patriotic reasons. One of the doctors
died during the experiments. The researchers discovered that an infected person would pass on the virus to a mosquito only after three days;
in turn, the mosquito would infect another person only after a period of
twelve days. In order to rule out miasmatism as a source of transmission,
the physicians made their volunteers sleep in dirty houses, wearing the
clothes of sick people. No one got infected from this treatment.[17]

Yellow fever proved lethal in one out of four cases. The first experiments had mostly produced only light cases of the disease. Encouraged
by these results, the researchers considered a vaccination, a deliberate
infection under medical supervision, but in the next series lethal cases
abounded. Gorgas, who was still not completely convinced of the mosquito theory, suggested a solution everyone else branded as absurd: the
eradication of the *Stegomyia* mosquito.[18] His plan also required a scientific procedure, based on zoology instead of medicine. Observations
showed that the mosquito rarely left the environment where it was born.
Water in uncovered receptacles was its most popular breeding place.
Gorgas started his campaign in February 1901. Employing the same
methods later used in Panama, he managed to eradicate yellow fever

within six months. His force fumed houses and stores, covered all containers filled with water, and dripped oil into ponds to keep the larvae from breathing.

When Gorgas arrived on the Isthmus, all scientific doubt regarding the mosquito theory had disappeared. Meanwhile, the British doctor Ronald Ross had proven with his work in India that the *Anopheles* mosquito was responsible for the transmission of malaria. In 1902 he received the Nobel Prize for his work. These discoveries notwithstanding, long-held beliefs prevailed in the minds of the Canal officials. Gorgas's superior Admiral Walker, head of the Canal Commission, "was not famous for a keen sense of humor, but the idea that there was anything dangerous in the bite of a mosquito stirred him to uncontrollable mirth."[19] In addition to this demurral, Gorgas had to deal with an entire jungle, not just a city, and not only with yellow fever but also with malaria, the infamous "Chagres fever."

He knew that malaria was the bigger challenge.[20] Yellow fever came and went suddenly; epidemics were always related to the presence of Caucasian people since Panamanians and West Indians were immune.[21] Those whites who survived were immune for the rest of their life, but other whites suffered a painful death. Malaria, on the other hand, was endemic on the Isthmus. Estimations based on hospital data suggest that three out of four of the inhabitants of Colon and Panama City were infected with the virus.[22] Malaria had a long-term negative effect on physical fitness and caused depressions—a considerable economic threat for the Canal construction. Gorgas's team found out that the *Anopheles* mosquito, as opposed to its striped compatriot, was able to travel longer distances and was not very particular about the water quality. The only realistic goal the sanitary officer could reach for was a reduction of the infection rate—the eradication of the mosquito was out of the question.

Gorgas started fuming the living quarters of infected persons. In the end, every house in the Canal Zone was treated as many as three times. In a short time, he used up an entire year's supply of insect powder in the United States. The infected laborers were completely shut off from their environment. In Washington, D.C., Gorgas's requests for money were not received well. In June 1905, at the climax of a yellow fever epidemic, the new commission chairman Theodore Shonts and the governor of the Canal Zone, Charles Magoon, demanded Gorgas's resignation. At that time, the army officer still had to report to the governor. Only after the abolishment of that position under the Goethals regime did he become a member of the commission (in spite of which the struggle for money continued). President Roosevelt refused to fire Gorgas and instead granted him greater powers on the Isthmus. Following the recommendations of the new chief engineer Stevens, he also agreed to increase the financial budget. While the previous annual budget for the sanitary work had amounted to $50,000, Gorgas now had $90,000 available for mosquito screens alone. In December 1905 the last case of yellow fever for the entire remainder of the construction period was registered. In June there had still been sixty-two cases, nineteen of which were fatal.[23] After decades of futile fights against yellow fever on the Isthmus, the problem was finally solved.

Gorgas also tackled malaria and the *Anopheles* mosquito. The sanitation workers dried out swamps, cut grass and bushes, and treated water surfaces with oil and larvicides. According to the chief sanitary officer, the quota of infected Canal laborers sank from 80 percent in 1906 to less than 10 percent in 1913.[24] Some illnesses, such as pneumonia, also decreased in case numbers, while others, such as tuberculosis, were on the rise. During the final year, the death rate for the American gold-roll employees sank to a record low of two per thousand, but the rate for the Caribbean workers was still four times as high.[25] According to official figures, 350 white Americans and more than 4,500 West Indians were killed by diseases, other illnesses, and accidents during the entire construction period. The historian Michael Conniff estimates that the latter number was in fact close to fifteen thousand which would mean that one tenth of the Caribbean workforce died on the Isthmus.[26]

Gorgas divided the Canal Zone into twenty-five sanitary districts, each headed by an inspector and equipped with rest areas and first-aid stations. The hospitals in Ancon and Colon, built by the French, were renovated and expanded. American and West Indian workers were placed on different floors. According to the Hay–Bunau-Varilla Treaty, the American government was also responsible for public health and sanitation in the Panamanian cities of Colon and Panama. Gorgas's agency built sewage canals, cleaned and paved streets, and oversaw the garbage disposal.[27] In 1908 the work was basically completed. The improved water quality helped to reduce the number of cases of typhoid fever and dysentery considerably. In total costs, the sanitation measures amounted to $20 million. About $6 million of this sum was spent on actual health services, which amounted to 1.5 percent of the entire Canal budget.[28] And yet sanitation efforts remained limited to the vital areas of the Canal Zone, often defined by race. To cover up this discrimination, malaria's persistence was blamed on the West Indians' presumed preference to live in the bush.[29] The sanitation workers took an excessive interest in the racial composition of the population. Beginning in 1910, the physician Herbert Clark, who took over the Canal Zone's medical laboratory after Gorgas's death, examined the skulls of dead Canal laborers and came up with results supporting his theory of the West Indians' mental inferiority.[30]

The Panama authors described Gorgas's battle against the insects in every detail, sometimes adding ironic comments. One writer pondered the question whether the holy water basin in front of the cathedral also offered breeding opportunities. "Perhaps holy water mosquitoes are innocuous," he concluded.[31] Critic Poultney Bigelow jeered at the Canal Zone buildings with their characteristic mosquito screens, which made them look "like a string of wire meat-boxes behind the screens of which our tame and timid officials collapse into rocking chairs, marveling at the progress of science!"[32] Others laughed at the order that any mosquito was to be reported to Gorgas's troops, with the intent that a sanitation worker could chloroform it and then arrest the convict in a test tube.[33] "She of the striped stockings and the shrill song," journalist Frederic Haskin tenderly called the dangerous *Stegomyia* female.[34]

These comments may have expressed a disbelief in the mosquito theory still lingering in the back of the authors' minds, but they also pointed to another aspect of the sanitation work. While Gorgas's efforts were primarily aimed at eradicating diseases, they also assumed the social function of supervision: "Preserving order and preserving health became synonymous."[35] A search described by the author Ira Bennett illustrates the drastic measures taken to track down possibly infected persons:

> Here was a man who was registered at a Panama Hotel. He was sick and some one feared he had yellow fever. When the authorities came to look him up, he had disappeared. The next day he was found in the streets intoxicated and suffering from yellow fever. He was taken to the hospital, where he died. Then they looked for his associates. Nobody seemed to know him. Finally it was heard that some of his countrymen frequented a certain bar room. Here again no one knew him, but several of them had heard him talking with an Italian. The Italians of the entire City of Panama were canvassed, and at last the man who had

talked with him was found, but the man knew him only slightly. How-ever, he did know that the man was acquainted with the watchman at a certain theater. This watchman was hunted down and was found to be ill with yellow fever himself. Then a little girl who frequented the thea-ter was found to have taken the disease. Every case was thus rigidly in-vestigated and all sources of infection run down.[36]

A number of other writers also printed a version of this story to demon-strate the success of disease control and did not seem to mind the almost brutal execution ("were canvassed," "was hunted down") of the search.[37] The commission came up with increasingly detailed orders, even after yellow fever had already been weeded out in the Canal Zone: special at-tention should be paid to rain water accumulating on machines that were no longer in use since "no precaution must be neglected at any time."[38] Gorgas reminded the managers at the hotels, restaurants, and mess halls that "table linen sufficiently stained or soiled to attract flies will not be allowed" in order to prevent infections.[39] The authors re-garded these precautions as part of the autocratic regime they had come to find and admire in the Canal Zone: "The methods of Panama were arbitrary, and had to be. They probably could not be enforced at all in a democratic community in ordinary times. The people would rebel against the severity of the regulations and against the incidental inva-sion of their privacy."[40]

For the Panama authors, Gorgas's success was first of all a triumph of American civilization. "Something very like a marvel has been accom-plished at Panama. A veritable valley of death has been converted into a land of health and comfort," commission secretary Bishop wrote.[41] Gorgas, "commander-in-chief of the forces of cleanliness and health," had achieved a miracle, the journalist Abbot noted, "changing the Isth-mus of Panama from a pest-hole into a spot as fit for human habitation as any spot on the globe."[42] The Panama authors usually devoted an entire chapter to the sanitation efforts. War metaphors such as Abbot's were not unusual, suggesting victory after a strenuous battle against dis-ease.[43] The writers emphasized that the death rate among the white Americans was eventually lower than in the United States. The Canal Zone became a model for the country as a whole, even though most of the workers were healthy young men. Gorgas received letters from gov-ernment officials, businessmen, and doctors all over the world asking him for the secret of his success. A senator from Maryland, for instance, wanted to protect his summerhouse on the Eastern Shore against mosquitoes. He "desires to know whether it would be practicable for him to take any individual measures to rid himself of the mosquitoes. He will be glad to furnish a map showing the location of the land and the situation of the house if same is desired," Chief Engineer Goethals wrote to Gorgas when he forwarded the request.[44] A cotton plantation from Arkansas, a railroad company from Brazil, and the government of Australia also turned to Gorgas with similar inquiries.[45]

According to the chief sanitary officer, the work achieved had implications reaching beyond the Canal project. Human civilization, he argued, had once originated in the tropics, but "the most vigorous and healthy races" eventually immigrated to the moderate zones. Deadly diseases had prevented a resettlement, "therefore the white man was permanently barred from building up any great civilization in these regions."[46] The successful sanitation work would enable Europeans and North Americans to return to the tropics in great numbers over the next few centuries, Gorgas predicted. An author for the *National Geographic Magazine* shared this vision: "New blood is needed in the tropics. The suns of centuries have burned out much of the initiative, the easy methods of gaining a livelihood have taken out much of the thrift, and the lazy ways of the tropics have eliminated much of the natural love of cleanliness of the people. New blood coming in may change these things to a very appreciable degree, and an even newer and better era of public health may ensue."[47]

Gorgas's theory, a racist perversion of the biblical expulsion from paradise, was closely linked to the social Darwinist doctrine of the expansionists. Turner's frontier was open again, and the "civilization" of the tropics would take only a matter of years. At the same time, Gorgas and other observers interpreted the Western colonies—and their transformation—as mirrors of their own rapidly changing societies. In contemporary art and travel literature, the "jungles and deserts" of foreign continents also "mapped the landscapes of the Western psyche," sometimes even in nostalgic ways.[48] On the Isthmus, the achievements of science and rigorous administration in the Canal Zone became a showcase for the United States—an interpretation to which the Panama authors would return again and again. "New blood"—or rather new methods of coping with the challenges of industrial society—was also needed at home, in the future America.

Passage and Network: Charting the Course of Empire

The Panama authors wrote about a range of topics regarding the work on the Isthmus, but naturally focused their attention on the seaway itself. "The Canal, after all, is the thing," journalist Willis Johnson believed.[49] The storytellers turned the Panama Canal into a symbol signifying far more than the economic benefits of trade it was expected to deliver. "No material work of man since the creation of the world has had so deep and widespread an influence upon the affairs of mankind in general as that which may calculably be expected to ensue from the achievement of the Panama Canal. The results will be seen in commercial, political, social, and even religious, effects," one author suggested.[50] Multidimensional interpretations such as these were not unusual for works of technology, especially if they were charged with metaphoric meanings. Despite this tradition and the historiographic attention it has deserved, the Panama Canal is curiously absent from most studies on the subject.

In the course of the nineteenth century, the building of bridges, canals, and railroads enabled Americans to reach their destinations more easily and faster. Connecting places, these engineering feats became symbols of the union of the young and expanding nation. Human technology triumphed over nature. The historian David Nye discusses and interprets this process using well-known examples—not the Panama Canal—in many of his works.[51] Nye concludes that most of these civic projects invoked a sense of collective identity—bordering on religious feelings—and "fuse practical goals with political and spiritual regeneration."[52] Depending on the specific case, the collective identity could refer to a community, region, nation, or even the world. Since federal investment began to play an important role only in the twentieth century, many of the works were financed by the respective states or private investors, including so-called joint-stock companies.[53] Their impact was often extended beyond the local or regional sphere. To some observers, the 363-mile-long Erie Canal, crossing the state of New York from Albany to Buffalo and completed in 1825, held a significance reaching far beyond the Eastern United States. Twice as long as any canal in Europe at the time, "it must have an important bearing upon the destinies of this nation, and eventually upon the whole world," a local newspaper commented.[54] In contrast to the endeavors of European monarchies, it was seen as the achievement of American democracy, "of the capabilities of a free people."[55] At the opening of the Brooklyn Bridge in 1883, Brooklyn mayor Seth Low exclaimed: "This great structure cannot be confined to the limits of local pride. The glory of it belongs to the race. . . . It is distinctly an American triumph. American genius designed it, American skill built it, and American workshops made it."[56] In France, the Saint-Simonians, a group of social visionaries influential in promoting the Suez Canal, also believed that the construction of railroads and canals was necessary to lift the public spirit and, by implication, build utopian communities reaching across the world.[57] The creation and interpretation of the Panama Canal was strongly influenced by this tradition, yet it must also be seen in its own context. The Canal came to symbolize two important and in many ways interrelated concepts: the passage and the network.

The Isthmus of Panama was the location where Christopher Columbus and other explorers had searched for the passage to India. On his fourth and last voyage from 1502 to 1504, Columbus had entered the mouth of the Chagres. In effect, the Isthmus was "the birthplace of American history"[58] and the Canal "the realization of an idea four centuries old,"[59] the Panama authors concluded. Its purpose was "to fulfil the great designs of Columbus and Cortez,"[60] the conqueror of Mexico who had allegedly suggested the building of a waterway to the Habsburg emperor Charles V. The myth of the "discoverers," the search for a passage to the Pacific Ocean, remained an important factor in the exploration of the American West. After the Louisiana Purchase, President Thomas Jefferson gathered an expedition force headed by Meriwether Lewis and William Clark, who were told to find "the most direct & practicable water communication across the continent for the purposes of commerce."[61] Encouraged by hearsay and the theory of symmetrical geography, Jefferson, himself a scientist, was hoping to find a navigable river flowing westward, located close to the source of the Missouri. He had pictured the western landscape as a kind of "garden," similar to his home state Virginia, and expected the Rocky Mountains to be of comparable size to the Appalachian Mountains, allowing for a water route to the West.[62] The Lewis and Clark Expedition would prove these preconceptions wrong, but the ideas of the passage and the garden, part of Jefferson's utopian imagination of a communal society, lingered on. For decades, Congressman Thomas Hart Benton argued for the development of the West to facilitate trade with Asian countries, later aided by New York merchant Asa Whitney, who proposed the building of a transcontinental railroad.[63]

The Panama Canal was seen as the true realization of the passage, the continuation of westward expansion, and, therefore, the affirmation of Turner's frontier thesis. Navy admiral Harry Harwood Rousseau, a principal engineer on the project and member of the third Canal Commission, saw the seaway as an inevitable consequence of manifest destiny: "Since the first hardy adventurers, pushing westward from their native shores, landed on the American coast, there has been no more doubt that this project would not, as an indispensable factor in the future of the American continent, ultimately materialize, than that those pioneers would not continue their westward journey overland from the North Atlantic coast to the Mississippi River, thence over broad plains and rugged mountains, and finally, as has long since been seen, reach the Pacific Ocean, carrying with them and leaving in their trail the energy and spirit that have developed and now maintain the American nation."[64]

Based on the concept of the passage, the building of the Canal—for many years a French endeavor, now made possible by the advances of science and technology—was interpreted as a distinctly American achievement, connecting the past and the future of the nation and the continent. And similar to earlier engineering works, the meaning of the Canal could be extended even further, encompassing the entire globe in the shape of a network.

The first railroad to the Pacific was eventually completed in 1869. Thrilled by this event as well as the opening of the Suez Canal in the same year and the first transatlantic telegraph cable realized only a few years earlier, the poet Walt Whitman praised the passage to India enthusiastically in his famous poem of the same title:

> Passage to India!
> Lo, soul, seest thou not God's purpose from the first?
> The earth to be spanned, connected by network,
> The races, neighbors, to marry and be given in marriage,
> The oceans to be cross'd, the distant brought near,
> The lands to be welded together.
>
> A worship new I sing,
> You captains, voyagers, explorers, yours,
> You engineers, you architects, machinists, yours,

> You, not for trade or transportation only,
> But in God's name, and for thy sake O soul.[65]

Whitman's dream was universal brotherhood, made possible by human passion and technological progress. The Panama authors expressed the same sense of exaltation regarding the waterway. It was "mankind's dream of the ages,"[66] an effort "to achieve at last the triumphant fulfillment of the world's age-long desire."[67] Abbot concluded that "in tearing away the most difficult barrier that nature has placed in the way of world-wide trade, acquaintance, friendship and peace, we have done a service to the cause of universal progress and civilization the worth of which the passage of time will never dim."[68] As literary historian Bill Brown points out, a "dialectic of the national and international" characterized the Canal interpretations, matching Roosevelt's "mandate from civilization."[69] The United States had taken on a special role in world affairs—not by means of its democratic tradition, as evoked in the building of the Erie Canal, but through the impetus of its expanding empire. Nationalism and internationalism were connected. Under the impression of the emerging war in Europe, the call for global friendship and peace was loudest at the Panama-Pacific International Exposition in San Francisco in 1915, even though the World's Fair shamelessly displayed the alleged superiority of the American nation and the white race: "Human endeavor has supplied no nobler motive for public rejoicing than the union of the Atlantic and Pacific Oceans. The Panama Canal has stirred and enlarged the imaginations of men as no other task has done, however enormous the conception, however huge the work. The Canal is one of the few achievements which may properly be called epoch-making. Its building is of such signal and far-reaching importance that it marks a point in history from which succeeding years and later progress will be counted. It is so variously significant that the future alone can determine the ways in which it will touch and modify the life of mankind."[70]

The welding-together of space, "the union of the Atlantic and Pacific Oceans," the earth "connected by network," as Whitman had put it, also had a modern, technological aspect to it. Industrialization, through steam-powered railroads and ships, had created a "new sense of distance."[71] While the ocean network promised fast and far-reaching access for purposes of trade, it also gave pivotal power to the one in charge. "Control of the sea, by maritime commerce and naval supremacy, means predominant influence in the world," Mahan had written, "because, however great the wealth product of the land, nothing facilitates the necessary exchanges as does the sea. The fundamental truth concerning the sea—perhaps we should rather say the water—is that it is Nature's great medium of communication."[72] Thomas R. Marshall, vice president under Woodrow Wilson, took a similar view: "I believe that the Panama Canal is destined to be the Marconi system of commerce."[73] Mahan and Marshall interpreted the seaway as a powerful, global *communication technology*, preceding twentieth-century inventions such as satellite navigation, mobile phones, and the Internet.[74] In a sense, they were paving the way for another expansion of the American empire, based not on physical coercion or superior engineering alone but on the exploitation of power relations embedded in networks. Historian Matt Matsuda detects a similar strategy in the French Canal project, destined to "make a 'French Pacific' through a masterstroke not of conquest or territorial expansion but by integrating, relaying, and connecting a set of disparate points into a tenuous web of empire."[75]

For the Panama authors, the Canal was a singular event in world history, representing both the past and the future America. Realizing Columbus's and Jefferson's dreams of the passage to India, it was seen as a uniquely American venture, the continuation and fulfillment of westward expansion, uniting the people in a common mission. The waterway proved American superiority, "forever a monument to the dauntless courage, infinite resourcefulness, ingenuity and administrative ability of the American people."[76] Destined to lead the world into the twentieth century, the United States would promote universal peace and friendship through the Canal. The concepts of the passage and the network represented the evolution of "a new Empire with open, expanding frontiers," a limitless America.[77] Unlike earlier engineering projects, the achievement of the Panama Canal occurred *outside* of the United States, literally paving the way for a globalization of American power in political and economic, as well as technological and moral, dimensions.

Pyramids on Broadway: Representing the Engineering Feat

The Panama Canal, almost fifty miles long, was a complex construction site, consisting of locks, cuts, dams, and a future lake. To keep their readers interested, the Panama authors had to choose for description those elements of the work that would convey an impression of the entire effort. How could the Canal itself be represented, how would it be constructed for the public? The most useful section for this kind of interpretation proved to be Culebra Cut. For a distance of eight miles, the Canal actually had to be cut out of the Panamanian rock. The ravine was eventually more than forty feet deep and three hundred feet wide at the bottom. Frequent slides threatened to detain the operation of the Canal even after its opening. One of these, the Cucaracha Slide, became a celebrity in its own right. "The work of months and years might be blotted out by an avalanche of earth or the toppling over of a small mountain of rock," Canal Commission secretary Bishop noted.[78] The cut was the ultimate challenge of the project and its biggest tourist attraction.[79] "Now stretches a man-made canyon across the backbone of the continent; now lies a channel for ships through the barrier; now is found what Columbus sought in vain—the gate through the west to the east. Men call it Culebra Cut," the author Frederick Haskin wrote, employing the rhetorical figure of anaphora to support his argument: "It is majestic. It is awful. It is the Canal."[80]

Flanked by the railway, the Culebra Cut was located in the southeast of the Canal Zone, stretching across the continental divide from Pedro Miguel Locks to the junction with the old Chagres River. As opposed to the locks and dams, there was little engineering ingenuity involved in the digging of the cut. Instead, it was "within the range of the comprehension of the ordinary person";[81] it symbolized the struggle against nature's forces and the achievement of the passage to India. The cut was also chosen as the background for the official seal of the Panama Canal, crafted by the famous jeweler Tiffany. The motive "consists of a shield, showing in base a Spanish galleon of the fifteenth century under full sail coming head on between two high banks, all purpure, the sky yellow with the glow of the sunset,"[82] invoking Columbus's "discovery" of America and his voyage to the Isthmus. The seal had the inscription "The Land Divided—The World United."[83] Images of technology were absent from the depiction. At the time of its design, it was still believed that a sea-level canal could be built.

The six locks with their tremendous steel gates, the 110 feet wide by 1,000 feet long concrete chambers, and the giant Gatun Dam and its spillway also caught the authors' attention, but not to the same degree.[84] Bullard tried to give an impression of the size of the dam, which was almost half a mile wide at its base, by calling it "a mountain range." The blocking of the Chagres created an artificial lake covering 164 square miles. "Never before has man dreamed of taking such liberties with nature, of making such sweeping changes in the geographical formation of a country."[85] Once again, phrasings such as these, mirroring the social upheavals of the age, may also have expressed an underlying fear that American engineering skills would not live up to their promise. One writer even unfolded an apocalyptic vision for the age of technological achievements: "Panama is a world-feature, from this on to the end of time, or until that native New Zealander shall sit on a broken arch of Brooklyn Bridge and contemplate the ruins of a civilization dead and turned to dust."[86]

The Canal attracted thousands of tourists to the Isthmus. It was believed that 85 percent of transit passengers came for the purpose of seeing the waterway, or "Canal spotting," as it was referred to. The number of these visitors climbed steadily during the final years of construction, from 15,790 in 1911 to 20,946 in 1912 and 18,972 in the first six months of 1913 alone, as reported by the *Canal Record*.[87] These masses could only be managed with strict organization: "The sightseeing business has therefore been systematized and its conduct is now a regular part of the work. There is no better way to see the Canal than the trips of the sightseeing train, and none that requires so little time. In any two consecutive week days it is possible to see the entire work. The train moves slowly through Culebra Cut, and about the locks and Gatun Dam, while the guide explains in clear and authoritative manner all phases of the work, and answers all questions."[88]

After the first half of the train ride, in Ancon, the tourists would listen to a lecture by the official guide William M. Baxter Jr., illustrated by models of the locks and dams and the Canal Zone that could be electrically illuminated at night. Baxter complained about the behavior of his listeners, some of whom would dress "as if for a trip through the jungle,"[89] and the quality of available information on the Canal: "The male fool is annoying only when he becomes excited. He has read a book, or perhaps two books, about the Canal on his way to the Isthmus. Books on Panama are probably no more inaccurate than books on Tibet; but there are more of them. And the inaccuracies are the most interesting points, therefore these lodge more firmly in the head of the fool."[90]

The Panama authors employed innumerable statistics and farfetched analogies to relate the size of the Canal work, the "eighth wonder of the world," to scales with which their readers at home were familiar.[91]

Figure 3. Pyramids along Broadway represent the "spoil" dug from the Panama Canal (reprinted from Willis John Abbot, *Panama and the Canal in Picture and Prose* [New York: Syndicate, 1913], 134).

The *Scientific American* noted that a wall reaching from New York to San Francisco could be built with the stones and sand from Panama, or sixty-three Cheops pyramids lined up on Broadway in New York from the Battery to Harlem.[92] The drawing of this scenario looked like a futuristic vision of the city (fig. 3). "In generations to come, the canal, like the skyscrapers of our cities, will be viewed as a manifestation of the building genius of the American people, just as the pyramids of Egypt are not remembered so much as the work of a given Rameses as a manifestation of the big building instinct of the entire race," Scott predicted.[93] Comparisons to the technical wonders of Manhattan—tunnels, subways, and bridges—abounded.[94] The new Grand Central Station was "a 'baby' Culebra Cut," one author mused.[95] The German novelist Bernhard Kellermann drew on the Canal as a model to imagine a transatlantic tunnel that would be built within fifteen years under the supervision of an American engineer and allow for a twenty-four-hour train ride from New Jersey to the Biscaya coast.[96] The science fiction novel became a bestseller.

Apart from its portrayal in the books and magazine articles by the Panama authors, the Canal made its way into popular culture. In the year of its opening and the preceding years, the musicians of Tin Pan Alley recorded songs such as "Where the Oceans Meet in Panama (That's Where I'll Meet You)," "Coaling Up in Colon Town," and "Sailing thru the Panama Canal." At least three marches, including "Hero of the Isthmus," were composed.[97] Adam Forepaugh & Sells

Figure 4. Historical advertisement for a modern fountain pen, 1914 (reprinted from *Town and Country*, May 16, 1914).

Brothers conceived the patriotic circus spectacle "Panama; or, The Portals of the Sea; or, The Stars and Stripes."[98] The Canal was also used in advertising. The Waterman Company praised the economical ink usage of its fountain pen by proclaiming "Great Time Savers: Waterman's Ideal Fountain pen and the Panama Canal" (fig. 4).[99]

All of these attempts at presentation and interpretation served to make sense of the Canal. Its meanings were shifting between simple, recognizable symbols—the digging of a big ditch, the gate to the West— and science fiction analogies of a future civilization, made possible by technology and engineering. Just like the tourists on the Isthmus depended on their guide, the middle-class readers and spectators at home needed the assistance of the Panama authors and other translators of culture in coping with the rapid, often frightening changes at the dawn of the modern age.

Moral Equivalent of War:
The Big Job and American Society

Searching for metaphors representing the Canal effort, the Panama authors noticed similarities with another heroic, physical, and manly endeavor: war. The sensual experience on the Isthmus already suggested the analogy: "You will see the same thick, black clouds of smoke," one author wrote. "Instead of the belching cannon, you will find a hundred times more deadly instrument in the giant dynamite blasts. The monster steam shovels, the great levellers and air-drillers are the weapons of warfare, and the opposing forces are the armies of man and Nature."[100] In the foreign territory, the perilous powers of the tropical environment found their match in the tools of American civilization—human labor and ingenuity—resulting in a "gigantic battle against floods and torrents, pestilence and swamps, tropical rivers, jungles and rock-ribbed mountains."[101] It was this fight against the *other*, the unknown enemy, that found a convincing expression in the war metaphor. The deadly diseases were part of this scenario. The author Abbot called Gorgas "commander-in-chief of the forces of cleanliness and health" and his efforts the "war of science upon sickness."[102]

To boost morale after the slow start of the project, Roosevelt, during his visit to the Canal Zone, compared the task before the Canal workers with the fighting of the Civil War, in which many of their fathers had participated. "I am weighing my words when I say that you here who do your work well in bringing to completion this great enterprise will stand exactly as the soldiers of a few, and only a few, of the most famous armies of all the nations stand in history."[103] The fact that the Canal construction—like most actual wars—was a limited "conflict" may also have contributed to the ease with which the comparison was used. "In view of these facts, shall we be stretching the point too far when we say that the conquest of the Isthmus of Panama is a feat of the arms of peace as brilliant and as difficult as any ever accomplished by the arms of war?" asked an author in the *Scientific American*.[104] When Goethals was put in charge of the project, the analogy became even more compelling. Although the Canal construction was not officially a project of the U.S. Army or the Army Corps of Engineers, as many contemporary observers believed, former soldiers suddenly played a crucial role in its execution. "I now consider I am commanding the Army of Panama, and that the enemy we are going to combat is the Culebra Cut and the locks and dams at both ends of the canal," Goethals remarked shortly after his appointment.[105]

The Canal had always played a role in actual military considerations, especially with regard to its fortification. The topic turned into a heated debate when American involvement in World War I became imminent.[106] In spite of this strategic role, the Panama authors as well as government officials never considered the Canal Zone a military state. Goethals was not a friend of army rituals and attire, and he emphasized the civil character of his regime on the Isthmus. "There will be no more militarism in the future than in the past," he reassured the workers. "Every man who does his duty will never have any cause to complain on account of militarism."[107] Nevertheless, the writers approvingly interpreted his rule as a dictatorship and discovered manifold parallels between the organization of an army and the mini-state erected in the Canal Zone, as is discussed in a later chapter. The war analogy was always used as an abstract concept, allowing the authors to give their readers a better sense of the spirit of the work and, going further, to turn the Canal project into a model for American society as a whole.

William James's (1842–1910) influential essay "The Moral Equivalent of War," published in 1910 shortly before his death, served as a basis for these arguments. The pragmatist philosopher, a former professor at Harvard, had been the teacher of many students who continued on to careers in politics and journalism, including Theodore Roosevelt and members of the New Intellectuals such as Walter Lippmann. Although an anti-imperialist opposing war and denouncing TR's call for the "strenuous life" as too vehement,[108] James still reached the conclusion that "martial virtues" resembling his former pupil's agenda for a moral rejuvenation of the country "are absolute and permanent goods."[109] The passion for war had to be replaced by something else, an equivalent: "If now—and this is my idea—there were, instead of military conscription a conscription of the whole youthful population to form for a certain number of years a part of the army, enlisted against *Nature*, the injustice would tend to be evened out, and numerous other benefits to the commonwealth would follow. The military ideals of hardihood and discipline would be wrought into the growing fibre of the people; no one would remain blind, as the luxurious classes now are blind, to man's real relations to the globe he lives on, and to the permanently solid and hard foundations of his higher life."[110]

James's suggestion of an army "enlisted against *Nature*" would be brought up again during the New Deal—for instance, with regard to the Civilian Conservation Corps[111]—but the building of the Panama Canal was an obvious demonstration of what the philosopher had in mind. Similar to the American workforce on the Isthmus, James's conscripts seemed exclusively male.[112] According to historian Athena Devlin, James believed that martial training caused "intense feelings" related to mystical or even hysterical experiences rooted in the subconscious, changing "the self-indulgent and effeminate 'insignificant individual' into a fighter."[113] A war-like effort such as the Panama Canal was the modern cure for a society threatened by excessive individualism and lacking manly resolve. It was precisely this image of the weakened, over-civilized element in American society that the Panama authors assigned to the French failure on the Isthmus.

The French project, abandoned in 1889, had left visible traces in Panama. Defunct machines, rusting in the swamps of the Isthmus, symbolized the tragic defeat of a nation overtaken by progress. Canal Commission secretary Bishop wrote: "The little locomotives and cars, almost toy-like in appearance when compared with those in use by the Americans, bore eloquent testimony to the irresistible onward march of mechanical invention. Time had retired them from active service as completely as if they had never existed, leaving them stranded as mere 'junk' along the wayside of progress. Covered with the softening mantle of vine and leaf and flower, and overshadowed by waving palms, they stood, in silent dignity, as the fitting monuments of a 'lost cause,' making a spectacle so eloquent with the sadness of failure, the pathos of defeat, that few beholders could contemplate it unmoved, and no Frenchman could look upon it with eyes undimmed."[114]

Depicting the romance of ruins, Bishop created the impression as if the French had retreated from Panama centuries ago, forgoing an entire civilization with them. "There is no richer digging in the ruins of ancient Rome or Pompeii than along the deserted route of the canal," an author for the *Scientific American* wrote.[115] The Americans' subconscious fear of their own failure may have played a part in inducing such sentimental observations.

But the authors went further. Most of them described the French efforts as an "opera-bouffe"[116] characterized by stupidity—supposedly, the Europeans had brought twenty thousand snowplows to Panama, "stored away in a territory where the average temperature is 110 degrees"![117]—as well as corruption and decadence. They showed little hesitance to repeat every national stereotype imaginable. "Champagne, especially, was comparatively so low in price that it 'flowed like water,' and other wines were to be had in scarcely less profusion and cheapness. The lack of a pure water supply was doubtless the moving cause for this abundance, which was justified on the ground of health preservation, but the consequences were as deplorable as they were inevitable. The ingredients for a genuine bacchanalian orgy being supplied, the orgy naturally followed," Bishop continued.[118] Panama author Haskin believed that "the whole course of the project was marred by an orgy of graft and corruption."[119] His colleague Abbot confirmed: "Wine, wassail and, I fear, women, were much in evidence during the hectic period of the French activities."[120] At the time, the waste of public or private money for alcohol, gambling, or luxury items—the French director-general had reportedly ordered the construction of a bathhouse worth $40,000—was denoted by the term "graft," a popular keyword in the writings of the muckrakers. Sometimes the debacle was linked to simple incompetence: "French clinical assistance, as it is called, has never been good; it is not good even in Paris, and much less so in the Provinces."[121] Another author compared the failed Canal project to Napoleon's retreat from Russia, since there was "much of the same exquisite French dash about the two enterprises."[122]

In contrast to these colorful accusations, the American engineers refrained from criticism and instead praised the French for what they had achieved. "Much that was of inestimable value had been learned from the French and from their experience, and that they builded well so far as they went is the consensus of opinion of all those who know," Goethals concluded. He assigned the French fiasco not to inadequate technology or health care but to "poor and maladministration."[123] In another comment, he stated that the French "would have carried the project through to completion had their financial situation been satisfactory."[124] Gorgas denied reports on the extravagant lifestyle of the French,[125] and Pepperman, who had conceived his book largely as a revisionist study, chided his fellow authors: "The general tendency seems to be to endeavor to cast additional luster on the achievement of our own engineers by ignoring the remarkable work done by the French. This is neither necessary nor just."[126]

Most of the authors did not pay any attention to the official evaluations by Goethals and others. They based their conclusions mainly on two sources from the French era: the British historian James Anthony Froude—who had apparently never visited Panama[127]—and the American Tracy Robinson, an employee of the Panama Railroad.[128] A witness of American acts of corruption as well, Robinson had also been cited as a source in Poultney Bigelow's articles. He was passionately opposed to alcohol consumption—"for it induces, by its derangement of the vital forces, every ill to which flesh is heir"—and similar vices, no matter who indulged in them, so his remarks on the "carnival of depravity" in Panama were the result of these personal convictions.[129] The Panama authors referred to him solely to denounce the French.

The writers chose to depict the French in accordance with Thorstein Veblen's observations of "conspicuous consumption" in the American upper classes. The description of European overcivilization reinforced Roosevelt's propaganda against the effeminate and morally corrupt capitalists at home. These projections were inherently gendered: words such as "sissy," "pussyfoot," or "cold feet" came into use around the turn of the century to sneer at unmanly behavior.[130] When Chief Engineer Stevens arrived on the Isthmus in 1905, he already had an explanation for the Canal crisis: "There are three diseases in Panama. They are yellow fever, malaria, and cold feet; and the greatest is cold feet."[131] The real objects of the authors' disdain were not the French in Panama but everyone in the United States who lacked the manliness and moral courage to revitalize the nation.

The fact that the French project had been privately funded was viewed as another reason for failure. "Most emphatically, if the desire for profit was to be the sole animating force the canal should never be built at all," wrote Abbot.[132] According to his fellow writer Haskin, the American Canal succeeded because "it was to be constructed not in the hope of making money, but, rather, as a great national and popular undertaking."[133] Again, the scolding matched the critique the Progressives voiced against big business in the United States and the individualistic "desire for profit." The authors associated corruption—which was indeed a problem for the French—not only with other immoral activities summarized under the term "graft," but also with private funding. This argument would play a large part in their utopian interpretation of the Canal Zone, where government control had eliminated the profit motive in favor of a collectivist state.

The French organizers of the Canal project would not have agreed with this view. Their venture was in fact heavily publicized as a "national and popular undertaking" in Haskin's sense, and not as a mere business opportunity. As in the case of the Suez Canal Company, the stocks of the Compagnie Universelle du Canal Interocéanique were distributed widely to the French public. Tens of thousands of small investors purchased shares, often mortgaging their real estate in order to raise the money.[134] This joint-stock company was interpreted as the "nation-state translated into capital."[135] This was precisely the reason why its bankruptcy catapulted the entire country into a moral crisis. While the American authors depicted French graft in order to point out what was wrong with the capitalist class in their own country, French critics such as Edouard Drumont mobilized a latent anti-Semitism in their search for an easy scapegoat. In both cases, meanings were assigned and shifted without much consideration of what had actually happened on the Isthmus.

Summing Up: The Dialectic Meanings of the Canal

In their success stories of the Canal, the Panama authors celebrated the unexpected turnaround of the events on the Isthmus. Once the work was on track, the "big job" was interpreted as a singular event in history, a miracle achieved by scientific and technological progress, social control, and manly resolve. The sanitation of the Canal Zone, containing the deadly diseases of yellow fever and malaria, became part of a national epic. In order to convince their readers at home, the authors had to construct meanings and metaphors along with the engineering feat; they had to translate the lessons of Panama into patterns that were recognizable and relevant at home. Culebra Cut became the synecdoche of the project, the digging of a ditch, the literal and long-sought passage to India. Like earlier public works, the seaway was assigned a significance far beyond its physical reach. According to the writers, the idea of the Canal was rooted in the American past, in the myth of discovery, and at the same time it paved the way into a utopian future. The Panama Canal was the arch of the American empire. In the new century, continuing expansion meant racial dominance—expressed in Gorgas's vision of the white man's "return" to the sanitized tropics—and control over "communication media" such as the oceans rather than old-style territorial conquest. American civilization expressed itself in a network of social power.

We encounter a threefold dichotomy in the authors' interpretations: past and future, national and international, war and peace. Although the Canal was described as a uniquely American achievement, it was also seen as a harbinger of global peace. This desired state, which may be characterized as "a perpetual and universal peace outside of history,"[136] was interpreted as the final order of empire, a web of power relations ensuring Western dominance. War was not needed any longer, and similar to the philosopher William James, the Panama authors turned the concept of war into an abstraction, the triumph of human skill over nature, disease, and the tropical *other*, the model for an efficient and autocratic state, and finally the recovery of manliness through Roosevelt's "strenuous life." On this level, the apparent contradictions of what the Canal stood for disappeared, and it came to represent all of these modernized concepts at once.

Many of the storytellers' constructs were rooted in the uncertainties within American society and the problems of industrialization. For demonstration purposes, the Panama authors imagined the French, who had pursued a national agenda not so different from the American objectives in Panama, as a mirror of the effeminate, morally corrupt capitalists at home. In their view, individualistic indulgence and the desire for profit had caused the French project to fail, while government control, a collective spirit, and manly virtues made the American triumph inevitable. Between the lines, however, the fascination with the French ruins on the Isthmus and other sentimental comments can also be read as expressions of self-doubt that the Canal and American society would ever live up to their promises.

3

The Engineered View

The Panama Canal in Pictures

In their efforts to interpret the building of the Panama Canal, policy-makers and popular authors relied both on texts and images. The dimensions of the project surpassed all prior achievements in the history of engineering. At the same time, it took shape in a foreign country, in the midst of the tropics. The task of pointing out the scale and significance of the Canal to their domestic audience could hardly be achieved through the written word alone. Compared to earlier projects such as the building of the Brooklyn Bridge, images could be produced and reproduced much more easily. As it happened, the waterway was "the first engineering enterprise occurring in the history of the kodak."[1] Politicians and journalists realized this advantage and employed the medium of photography and its contemporary derivatives (such as stereographs) as their tools—a strategy that proved crucial for the success of the project but that has received little attention from scholars.[2]

A flood of images from the Isthmus made its way to the United States. The most famous of these shows President Theodore Roosevelt operating a steam shovel during his visit to Panama. The government employed an official photographer, Ernest Hallen, to capture the building of the Canal in thousands of photographs. Newspapers, magazines, and publishing houses used his images to illustrate their stories and books on the project. News syndicates sent their own photo reporters to the Canal Zone, whose shots ended up in daily papers, in collections of stereographs, and on postcards. The artist Joseph Pennell depicted a few dozen scenes of the construction work in his lithographs. They went through several printings within one year.

Constructing Images: Problems of Interpretation

Around the beginning of the twentieth century, photography was still a relatively young medium, and its techniques and uses differed in many ways from today's established formats (which, under the impact of digital photography, have once again begun to change dramatically). This is one of the reasons why the analysis of early photography requires special caution. Many important variations, such as stereographs, have completely passed out of use. For many decades, these twin photographs, which produced an interesting three-dimensional effect when viewed through a special apparatus, had been an extremely popular form of visual entertainment in middle-class homes. Another difficulty for the historian lies in the fact that the reception of images can be studied only through other media, usually texts, and such texts are hard to find. "Eyewitness" descriptions of how an image unfolds its visual power, such as Oliver Wendell Holmes's famous essay "The Stereoscope and the Stereograph," are rare.[3]

There were two reasons why the appeal of photographs reached a new peak at the time of the building of the Panama Canal. First, new technologies allowed for cheaper manufacturing and printing of images. Second, what the images offered to display was precisely the same modern world of which the ascent of the medium was part. Domestic environments were transformed under the impact of industrialization, and foreign places and people were incorporated into the American empire. Photographs served to document this process. Their precision and presumed realism gave them a compelling advantage over traditional media and art forms. At the same time, efforts were made to position photography within the arts, to define and understand it solely through aesthetic categories, and to consider individual photographers as artists.[4] In her path-breaking essay "Photography's Discursive Spaces" on American landscape photography in the nineteenth century, Rosalind Krauss convincingly argues that an analysis based on photography as an aesthetic art form, designated for exhibition in a museum or gallery, greatly misrepresents and underestimates its actual impact.[5]

The historian cannot judge images by their alleged status as documents of "reality" or by their aesthetic appeal alone. Photographs, like texts, are constructions of meanings. They become part of discourses related to whatever they are displaying (or not displaying). "Even in the 1890s, inconsistency and confusion existed concerning photography as a hybrid of art and technology, but these obstacles did not prevent its use as a tool for science, an agent of social change, and a replica of things not present," historian Julie Brown postulates. In this sense, photography is "a cultural practice, a distinctive way of constructing, producing, and using images."[6] There is no fundamental difference between the functioning of written texts and images. Like paintings, photographs are interpretations of their subject rather than mirrors of reality. It seems feasible, therefore, to adapt methods used in art history, such as Erwin Panofsky's iconographic approach, to the analysis of individual or groups of photographs.[7] This kind of analysis positions aesthetic judgments within a social and cultural context. How is the image composed and what does it show? Which motifs and symbols are displayed? What intrinsic meanings, intended or unintended by the artist/photographer, are implied? When no additional information by the publisher, the photographer, or readers/recipients is available, a deductive approach such as Panofsky's is often the only way to assess the historical significance of an image.

I also acknowledge a method of studying images that does not consider photographs individually, or in any relationship to a particular author/photographer, but that views them as constituents of a vast visual system of representation, similar to Foucault's notion of the archive.[8] Visualization and its deviant, observation, thus function as manifestations of discourses.[9] This approach focuses on the contest for the meanings of images, the control of their production and dissemination, and the attempt at dominance over their subjects by establishing a photographic *gaze* (which, for instance, looks down on immigrants or exotic people).

There are thousands of photos of the building of the Panama Canal, and thousands alone were taken by the official photographer, Ernest Hallen. It would be problematic to consider Hallen the sole "author" of these images. Little personal information about him is known. It is possible, however, to study the circumstances under which these

photos were created. The U.S. government had a much greater influence on the distribution of images than on textual information regarding the Canal. Hallen's superiors were engineers who wished the day-to-
day progress of the work to be recorded, but his photographs were far
more than technical documents. They were sold to newspapers and
magazines and ended up in the books of the Panama authors. His and
other photographers' Canal images were placed and interpreted in
a context, such as the private stereograph collection of an American
middle-class family. I discuss individual photos in such contexts, and I
also address the broader perspective (or archive, if you will) created by
these Canal images. In her essay, Krauss borrows the term "view" from
the nineteenth-century survey photographers with whom she is concerned: "View addresses a notion of authorship in which the natural
phenomenon, the point of interest, rises up to confront the viewer, seemingly without the mediation of an individual recorder or artist, leaving
'authorship' of the views to their publishers rather than the operators (as
they were called) who took the pictures."[10] The government, the photographers, and the Panama authors constructed for American observers at
home an official, *engineered* view of the Panama Canal.

World Visions: The Evolution of Photography

Throughout the latter decades of the nineteenth century, technological
advances helped to make both production and publication of photographs easier and less expensive and therefore enabled the medium
to reach a mass audience. In the 1880s the gelatin-coated glass plate (or
"dry plate") was introduced and proved to be far less perishable than
its predecessor, the "wet plate," making more feasible the use of professional photography for short-term assignments such as news reporting
or for travel purposes. It remained a common technique even after
World War I.[11] In addition, the roll of film emerged as an alternative
that was easy to handle for amateurs as well. The first successful roll-film
camera, the Kodak, was introduced in 1888 by the Eastman Company.
It was advertised with the slogan "You press the button, we do the rest."[12]
The camera cost twenty-five dollars, which was still expensive but af-
fordable for many middle-class households who desired the new product. By 1896 more than 100,000 Kodak cameras had been sold world-
wide.[13] A popular roll-film camera used by professional reporters was

the Graflex, introduced in 1898, with its characteristic boxy look. The focus could be adjusted by looking down into a hood-shaped viewfinder. In order to show photographs to a larger audience, lantern slides were made of images and displayed with a projector ("magic lantern"). Although these slides were thick, heavy, and fragile, they were frequently used for lectures and presentations to expert audiences.[14]

Publishing photos had been a tedious and expensive process and was often combined with the use of more traditional artistic modes of reproduction such as lithography and photogravure. For the latter, a photographic negative was placed on a steel-faced copper plate, which was then etched in different depths proportionate to the tone of the image.[15] In the 1880s, the halftone process revolutionized the reproduction of photos in print. The mechanized procedure transferred the tonal graduation of photos onto the page by representing them in a matrix of dots and lines. From then on, photos were increasingly regarded as a necessary supplement to texts, even though modern photojournalism with multiple-page photo sequences in large formats and in color would be established only in the 1930s by magazines such as *Look* and *Life*.

The halftone process gave birth to the era of yellow journalism dominated by the publishing moguls William Randolph Hearst and Joseph Pulitzer. Supported by the evolution of photographic technology, it helped establish the profession of the news photographer. In contrast to articles by newspaper and magazine authors, most of their work was not credited.[16] The exceptions were war photographers such as James H. ("Jimmy") Hare, a British citizen who immigrated to the United States and became famous for capturing the live action of armed conflicts with his camera.[17] In general, however, news photography was not viewed as a genuine craft, and sometimes photos were retouched or painted over by graphic artists prior to publication.[18] The anonymity of the news photographers was also due to the fact that many of them worked for syndicates such as Underwood & Underwood, Keystone View, or H. C. White, which sold their images to newspapers and magazines such as *Harper's Weekly* and *Scientific American*. The same photos were used in illustrated books or published as stereographic sets. These companies photographed every major news event of the time, ranging from local accidents to global spectacles such as the Spanish-American War and the San Francisco earthquake in 1906.

Many of these images were literally "news" to people—things they had never seen before. Surveying and conquering unknown territories and encountering "exotic" indigenous people, whether in reality or through a medium, could be a disturbing experience, but the successful mastery of such confrontations with the mysterious, "awful," or allegedly supernatural resulted in feelings of empowerment and pleasure. This state or sequence of negative and positive emotions, described by the term "sublime," was an archetypical process of the enlightenment. It could be cut short and accelerated by experiencing the moment of "thrill" for a short time only, or from a safe position, or by looking at a painting or photograph. Traditionally, the experience of the sublime was triggered by an encounter with a natural wonder such as a waterfall or a canyon. During the nineteenth century, when "the technological wonders of American civilization such as canals, bridges, and trains caused waves of excitement," the focus of the sublime shifted from nature to culture, as Jürgen Martschukat observes, "from the natural spectacle to its technological appropriation."[19] On his encounter with Culebra Cut, Panama author John Foster Fraser noted: "[T]here is something dramatic, majestic, and occasionally terrible in it all."[20] In the twenty-first century, the sublime, in its postmodern variant, has entered completely artificial worlds such as cyberspace.

In the history of the media, we find numerous strategies to facilitate the experience of the sublime, to appropriate the unknown and guarantee the mildness of the horror, from the imitation of reality in today's video games to the aesthetic program of nineteenth-century landscape painting. Typically, American artists depicted their western landscapes from heights, in panoramas, implying the safety and superiority of the observer. The spectator's position was either identical with the position of the painter, or the spectator was included in the picture as an onlooker, either for scale effects or as a sign of the human presence in nature. Usually, Native Americans were excluded from these images. As Albert Boime notes, "the panoramic prospect becomes a metonymic image, that is, it embodies, like a microcosm, the social and political character of the land."[21] At the same time, it creates the illusion of the sublime as a "shared emotion," a collective experience rather than a function of the individual.[22]

A genre that adopted some of the characteristics of landscape painting was survey photography. Government agencies assigned photographers to their missions exploring the new territories in the West. The images produced on these trips served different, seemingly unrelated purposes. On the one hand, they were taken to document the topographical characteristics of the country and to record the progress made on infrastructure projects in these areas. For this purpose, they were addressed to a small circle of engineers and other experts. On the other hand, the images were instruments to justify expenses. Members of Congress who approved the budgets for expeditions and public works rarely appreciated geological details. Instead, they shared the general public's aesthetic tastes and fascination with the sublime. Photographers faced a difficult task: they had to please different audiences.[23]

Among the most prominent government photographers were Walter J. Lubken (1881–1960), who worked for the U.S. Reclamation Service and made images of irrigation projects, roads, bridges, and other public works, and William Henry Jackson (1843–1942), once a self-employed photographic pioneer who was asked to accompany the F. V. Hayden Survey to the Yellowstone area. A bound volume of Jackson's images was presented to members of Congress who then introduced a bill to turn Yellowstone into the first national park.[24] Jackson later joined the Detroit Photographic Company, added his own images to the firm's inventory, and became its president.[25]

Timothy H. O'Sullivan (1840–1882) was one of the most successful early survey photographers. An Irish immigrant, O'Sullivan began his career in the studio of civil war photographer Matthew Brady. From 1867 to 1869, he accompanied Clarence King, a young civilian scientist, on the nation's first major survey of the western territories, which also led him to Shoshone Falls in Idaho. A few years later he returned on a second expedition, this time led by army officer George M. Wheeler. O'Sullivan commented on the impressive landscape, noting that "below the falls, one may obtain a bird's eye view of one of the most sublime of Rocky Mountain scenes."[26] The elevated perspective popular in landscape painting was a natural choice for him. It was also a position granting the power of interpretation, for which Mary Louise Pratt has coined the term "monarch-of-all-I-survey scene." Characteristic of many of the imperialistic projects of the age, this position could become the pretext for intervention and mastery. "In the end, the act of discovery itself, for which all the untold lives were sacrificed and miseries endured, consisted of what European culture counts as a purely passive experience — that of seeing."[27]

It was O'Sullivan who accompanied the Darien expedition in 1870 to the Isthmus as a hired photographer.[28] Only three months after the grand opening of the Suez Canal, Commander Thomas O. Selfridge and about a hundred men arrived in Panama to search for a possible canal route. The earlier expeditions into the western territories and the construction of gigantic public works of technology, such as the Brooklyn Bridge on a national level and the Suez Canal in the international arena, had set the stage for this optimistic venture. Despite the existence of the railroad, the territory of the Isthmus—or Darien, as it was called at the time—was still largely unexplored. The San Blas route tested by the expedition proved unsuitable for a canal, Selfridge concluded. Nevertheless, the project provided new information on the country and its topography and was viewed as a success. Further expeditions took place in 1871 and 1873. On the second venture, John Moran served as the official photographer. Of the photographs that survived from these trips, it is unclear when exactly and by whom they were made, but it seems probable that it was O'Sullivan.[29] Most of the thirty captioned but uncredited prints held at the Library of Congress show tropical landscapes and village scenes with Native people. Images such as "Great Falls, Limon River" clearly resemble photographic motifs from the American West. It could also be argued that the photo of a natural arch at Cupica Bay alludes to an almost archetypical image of the American landscape, Thomas Jefferson's Natural Bridge in his home state Virginia.[30]

The photographer tried to capture well-known geological formations and mix them with more "exotic" shots of a mango tree or the huts and boats of the Natives. Apparently his objective was to incorporate the American landscape into the foreign territory, creating sublime images at once familiar and foreign to the observers at home. References to the actual purpose of the expedition appear in two captions: The photo "The Terminus of the Proposed Canal, Limon Bay," depicting the shore near Colon, helps the viewer to imagine how technology would transform the natural landscape. "Limon Bay, Low Tide" draws attention to an engineering aspect, the tide management of a sea-level canal. As in earlier surveys of the American West, these photos were probably intended to document the expedition for both technical and publicity purposes. The broader public had access to them through stereographs. At the time, these images of the Isthmus were "probably the most complete extant coverage of any Central American area."[31] The survey view bridged the gap between the unknown and the familiar and prepared the American middle class for expansion into foreign countries.

The rise of photography coincided with the rise of technology in general, of which it was a part. During the second half of the nineteenth century, photographers increasingly focused on the wonders of progress such as bridges, dams, railroads, and other public works, as well as the epochal world's fairs. The first of these in the United States was held in Philadelphia in 1876. These expositions were "show cases," and they employed visual media—traditional art forms as well as their modern counterparts—to explain progress. They were subjects of photography as well as places to exhibit images of the times.

Typically, the photographs of earlier technological projects tended to focus on groups of people, usually dignitaries, posing in front of the work while still in progress or on a celebratory occasion such as its completion or opening. These images mimicked established formats of portraiture and put less emphasis on the project itself.[32] Gradually, the depiction began to focus more on the sublime aspects of these endeavors. For the illustration of railroads and other infrastructure projects in the American West, landscape painting and survey photography had already supplied the visual language. As photography became widespread and the United States empire expanded into other parts of the world, this American "vision" no longer had any limits. As an evolving technology, photography reflected the rapid changes of the age. It became an instrument of explaining these changes, a negotiator between the optimism and fear felt by many middle-class Americans as they faced industrialization and expansion. The images of the Panama Canal played an important role in this process.

There were different ways to employ photographs as witnesses of modernization. The genre of documentary or social photography evolving in the American cities had a strong impact on the further development of the medium and deserves a brief discussion. Around the turn of the century, muckraking journalists and social workers who tried to expose the ills of urban life viewed photography as an unfiltered, authentic medium alerting newspaper readers to take notice of the hidden "details" of their environment. In contrast to the total shots and elevated vistas of the landscape photographers, these reporters often captured their subject matter from mid-range and close-up perspectives. The Danish-born Jacob Riis (1849–1914) employed photographs to document and authenticate his reports. In his major works, *How the Other Half Lives* (1890) and *The Children of the Poor* (1898), the self-taught newspaper reporter focused on the lives of urban immigrant families. The sociologist Lewis Hine (1874–1940) took images of immigrants arriving on Ellis Island and worked as a photographer for the National Child Labor

Committee. He went into factories, sweatshops, and mines with his camera. In the 1920s he completed the cycle *Men at Work*, an examination of the relationship between men and machines that has become a classic work of modern photography.

Riis was a friend and biographer of Theodore Roosevelt.[33] Many of the social photographers were associated with the Progressive movement. As in the case of survey photography, their images were used for more than one purpose. While they may have drawn the observers' attention to poverty and suffering in the big cities, they also helped authorities to document and eventually control disruptive social forces.[34] In terms of the composition of images, they proved more influential than the panoramic views of the landscape photographers. Their focus on the individual and the locale took a new shape in the 1920s, when they discovered the confined spaces of the corporation to illustrate the new relationship between machines and people. It was also the documentary work of Riis and Hine that paved the way for photographers such as Margaret Bourke-White and Walker Evans (who were increasingly regarded as artists). Their status as icons of photography may be one of the reasons why the work of photographers such as Ernest Hallen on the Isthmus has received little scholarly attention.

Photography was not a uniform medium: At the beginning of the twentieth century, stereographs and postcards were important variations and mass media in their own right. Stereographs were viewed through a special apparatus, the stereoscope, which had been in use in upper middle-class households for many decades.[35] In his famous essay written in 1859, Oliver Wendell Holmes discussed the impact of the new medium on human perception and called it "the card of introduction to make all mankind acquaintances."[36] By 1900 distribution channels such as mail order and door-to-door sales had helped to establish stereography as a popular leisure activity even in less wealthy households.[37] Photographs and postcards were often shared and looked at in groups (for instance, in a family's living room), but the viewing of a stereograph was essentially an individual experience. All other visual context shut off, the eyes were completely focused on the image, a setup Krauss calls "tunnel vision."[38] The "stereo" effect of three dimensions is stunning even for today's observers—probably because we have all but forgotten about the medium—and must have caused contemporary consumers to hold their breath. In terms of reception, stereography combined characteristics of photography and cinematography. The spatial viewing experience was achieved by looking at two similar images simultaneously. They showed the same object from slightly different positions. The stereophotographs were usually taken with a twin-lens camera. The stereo effect was drastically improved if the image had strong contrasts between foreground and background. This is why photographers often used compositional features, such as a human figure in the foreground, to suggest depth.[39] When the stereograph is viewed as a regular photograph, the special composition of the image often makes no sense.

After the turn of the century, stereographs were typically sold in sets that shared a common theme. The topics reflected the age of industrialization and expansion. Its novelties could be viewed from the safety of one's home, for purposes of entertainment or education. "Exotic" countries and their people were frequent subjects of stereographic sets, providing "a learning experience to comfortably domesticate that which was new or potentially threatening."[40] Stereographs of the evolving U.S. empire in foreign places such as Hawaii, Puerto Rico, or the Philippines were particularly popular.

Stereographs were sold by companies such as Underwood & Underwood and Keystone View, which employed full-time photographers. The brothers Elmer and Bert Underwood founded their firm in 1880 in Ottawa and got started with door-to-door sales of other publishers' images. The company expanded quickly and moved its headquarters to New York City in 1891. By then the brothers produced their own photographs and sold them to newspapers and magazines. Underwood & Underwood's freelancers took thousands of images each day. By 1910 news photography had become the company's most profitable division, and it started to sell its stereographic negatives to the competitor Keystone. In 1920 the production of stereographs was discontinued, and Keystone purchased the remaining business. It had already acquired the business of H. C. White in 1915.[41] Many images now assigned to Keystone had in fact been taken by its former rivals.[42] Founded by B. L. Singley, an amateur photographer, the Keystone View Company of Meadville, Pennsylvania, had set up an educational department that arranged stereographs into boxed sets. "See the world through Keystone" became the company's slogan. The sets could be ordered by mail, and cereal producers included free stereographs in their packages. Other customers included schools and libraries. The huge number of images served "to produce a comprehensive yet fanatically detailed vision of the world."[43] The editorial board of the educational division was headed by

Charles W. Eliot, a former president of Harvard University. The board was responsible for the detailed captions printed in guidebooks or on the back of the cards, which provided interpretations of the images, turning them into "organized experiences."[44]

Very often, stereographs were the only available distribution channel for the work of photographers such as Timothy O'Sullivan.[45] Before the construction of the Panama Canal began, O'Sullivan's Darien views, published as stereographs, were an important source of reference. When the Canal building was underway, the stereograph syndicates produced images in large numbers. Echoing the publication rush prior to the Canal opening, the stereograph output reached its maximum level during the final years of construction. William Culp Darrah lists a 36-card collection by Underwood & Underwood published in 1909 and another 45-card collection that came out in 1912. In the same year, Keystone View offered a 100-card box.[46]

In the 1900s, when the Panama Canal was built, stereographs were not the only visual medium to reach a mass audience. Picture postcards rivaled stereographs in their popular appeal.[47] They were first sold commercially at the Columbian Exposition in Chicago in 1893 as souvenir sets. In the following years, postcards remained less common than in Europe, but by 1905 Americans displayed an equally strong interest in collecting them. In 1906 more than 770 million postcards were sent through U.S. mail—many of them produced in German factories—which is an indicator of how widely their images were disseminated. Many images, which could be photographs or drawings, were printed a few thousand times. The building of the Panama Canal and the Panama-Pacific Exposition in San Francisco were prominent subjects. George and Dorothy Miller call the period from 1898 to 1914 the "postcard era."

The popularity of stereographs declined in the 1920s, as new competitors had arrived. Movies provided an enhanced visual experience, and radio became the most common choice for home entertainment. Postcard sales had already dropped rapidly in the 1910s. Rising postage costs and a lack of paper are named among the contributing factors.[48] The brevity of the postcard fad underscores the impulsiveness of the public's demand for new images. Driven by curiosity and the need to make sense of the changing environment, people shifted their attention to whatever media were at hand, depending on cost and availability. Images spurred the interest of middle-class Americans in the modern age and the changes it involved, and served to still this appetite at the same time. Visual media gave them an impression of the foreign world that progress and expansion had created around them: the American empire.

On Top of the Steam Shovel: TR's Imperial Body

Theodore Roosevelt's visit to the Canal Zone in 1906 turned around the public perception of the project from an ill-fated venture in the jungle to a modern success story of American masculinity and engineering. It comes as no surprise that photographs played a crucial role in his mission. Unlike any other president before him, TR used the media to establish the image that he had created for himself: "No American politician was more savvy about the press and photography than Teddy Roosevelt, who eagerly acceded to those wishing to record his likeness."[49] He knew that the effect of widely circulated photographs went far beyond purposes of documentation. As credible evidence, untampered by interpreters such as journalists, images could be employed to represent political achievements and visions, just like they had been a key factor in the construction of his own public personality. Numerous portraits of his participation in the Spanish-American War had been made. Before embarking on his trip to the Isthmus, TR decided to turn the visit into a campaign and gave instructions to Joseph Bucklin Bishop, the Canal Commission secretary and designated historian of the enterprise: "'I want pictures of everything and every place I am to see,' he told me, 'pictures that will show something, and mean something.'"[50]

Roosevelt's brief moment atop the ninety-five-ton steam shovel was a classic "photo op." It was wisely chosen for the very beginning of his tour of the Canal Zone. Later in the day, the rain season and the construction sites left their marks on the president, who appeared "wet through and spattered with mud."[51] Judging from the number of published and archived shots, all the relevant news syndicates were present to capture the scene. The most influential versions, in the following called A and B, were taken by Underwood & Underwood:[52] The shovel is in the center of the image with TR, wearing a Panama hat, at the controls. While the president sits tightly and appears not to be completely at ease with the situation, this impression is more than outweighed by the composition of the photograph. His white suit sticks out in stark contrast to the black, soot-drenched steel machine.[53] Instead of focusing on the center of the image, the eye of the observer can also follow a diagonal line from lower left to upper right. This line is a natural element of the composition and crosses TR's figure halfway to the top. It is impossible to escape his image. Both versions include additional people: a laborer in the upper right-hand corner holding a rope and a background group of workers in the lower right-hand corner. In version B, another laborer is standing in the lower left-hand corner with his back turned toward the camera, looking up to Roosevelt. All featured persons are white workers. Framing TR, the two figures on the diagonal line make version B particularly effective (fig. 5). When viewed through one of the contemporary stereoscopes that the Print & Photographs Division at the Library of Congress provides, it becomes clear that they contribute to an impressive depth effect, which literally makes Roosevelt in his "cockpit" come alive. They also represent the physical, everyday effort of the work, whereas TR's position at the lever, facing the photographer, symbolizes the abstract, centralized exertion of power. Manifestations of gender and race are inscribed into the photograph. Through his pose and attire, Roosevelt conveyed powerful images of manliness and whiteness to the audience back home.

For the contemporary observer, the image summarized the purpose of Roosevelt's trip. He had come to Panama to evaluate the situation and assure the public that the Canal project was under control und would live up to its promise. The impact of the photograph was calculated and its setup carefully constructed. TR's white figure became the embodiment of the new U.S. empire: "This body emerges when American technology, as epitomized by the Panama Canal, appears as the new mechanical mode of American international triumph, when Roosevelt is famously photographed not on top of a horse but sitting at the controls of the Bucyrus shovel at Pedro Miguel, the startingly [*sic*] white American in control of, but miniaturized by the gargantuan, dark prosthetic machine."[54] Even though the Canal project and its technological signifier, the steam shovel, loomed larger than the human body, Roosevelt's image illustrated that they could be tamed by the imperial mind. In 1912 the artist Joseph Pennell quoted a Canal laborer who thought that the steam shovel "would look just like Teddy if it only had glasses."[55]

After his return from Panama, Roosevelt delivered his message to Congress on December 17, 1906. According to Bishop, it was the first document handed to the members in print rather than script. It was produced in such a way that each double page consisted of text on the left and a photograph on the right-hand page, which has remained the

Figure 5. President Theodore Roosevelt on top of a steam shovel during his visit to the Canal Zone, 1906 (reprinted with permission from Keystone-Mast Collection, UCR/California Museum of Photography, University of California, Riverside).

preferred placement for advertisements in newspapers and magazines up to today. In addition, it featured a double-page panorama photo.[56] "This was the first illustrated message ever transmitted to Congress and its appearance in the Senate caused a feeling of consternation in that august body, whose members looked upon it as that abhorrent thing called 'an innovation,' a breach of tradition amounting almost to treason," Bishop wrote.[57] The Canal Commission secretary added that the reception of the illustrated message in the House was enthusiastic, and several editions were ordered. After a while, even the Senate changed its mind and made an order of ten thousand copies. They were widely circulated in the United States as well as in Europe.[58] Roosevelt had turned his trip to Panama into a public relations coup.

The Official Photographer:
Ernest Hallen's Panama Canal

At the time of TR's visit to the Canal Zone, the government did not employ an official photographer. The earliest reference to such an office was made by John Findley Wallace, the first chief engineer on the Isthmus. In February 1905, he wrote to his superiors that he had "tried two photographers; the first one failed on account of the quality of the work, and the second one got 'cold feet' about two weeks ago and left."[59] Then, for the brief period of three months, Franz Biedermann, a native of Austria, held the position of official photographer. According to the personnel records, Biedermann was assigned to the general quartermaster and began his work on June 5, 1905. A few weeks later, he was discharged.[60] Biedermann once requested a fan for his workplace "to reduce the temperature of the baths and plate process";[61] otherwise, there remains very little documentation of his work. He was let go when Wallace's successor as chief engineer, John F. Stevens, decided that there was not much for a photographer to do at that point, at least not until the progress of the work would be noticeable. So far, it had amounted to "practically nothing," Stevens wrote to commission chairman Shonts. The office, which Stevens viewed as "a waste of money," was abolished on October 15, 1905.[62]

In the spring of 1906, James R. Mann, a member of the House of Representatives, suggested to Secretary of War Taft that the Canal Commission employ an official photographer, "as I never see any pictures of the Panama Canal or the Panama country in any of the daily papers."[63] He even suggested a specific person from Brooklyn—not Hallen—for the job. In his reply, Stevens agreed that the expenses for such a position were indeed justified now that preparations were completed and the actual construction of the Canal was about to begin. He stated that he had even included working rooms for a photographer in a new administration building.[64] At first, Stevens had no one particular in mind for the job, but in a cablegram almost five months later—shortly before TR's visit—he wrote that he had "a competent photographer," which by the time must have meant Hallen.[65] When Roosevelt took his trip to the Canal Zone, Hallen had already begun his work as a freelancer for the government. A memorandum lists various charges for plates and prints, starting on August 28, 1906.[66] According to an accompanying letter, Roosevelt's visit kept Hallen busy: "A large portion of the work will be found to have been done under Mr. Bishop's orders at the time of the President's visit to the Isthmus."[67]

Among the engineers, the photographic work done on the Isthmus remained controversial. One of them complained about the images Hallen took on two particular days and refused to pay for them. He thought they were "of absolutely no value as they were evidently taken for pictorial effect and not for showing progress of our work."[68] In a letter to Goethals, who had replaced Stevens as chief engineer in the spring of 1907, supervisory engineer W. L. Sibert made the proposal that, once again, a photographer be fully employed by the government. "It is thought that there will be enough photographic work to keep a man busy all the time," he wrote. The government would be able to "control the issue of every print." Sibert also felt that the army engineers had been overcharged for copies of existing photographs. Hallen was a natural candidate for the position. "It is suggested that the present photographer be asked at what price he would engage to devote all of his time to the Government of the United States without the privilege of selling any prints whatever of the negatives taken for the Government."[69]

Accordingly, Hallen became a full-time employee effective August 1, 1907. Until then, he had charged $5 in gold a day—but only for a maximum of five days a month—plus 25 cents for each plate and 35 cents per print.[70] His service record card states that Hallen was from New Jersey and thirty-two years old at the time of his appointment. He was given a salary of $175 per month. Ernest Hallen (1875–1947), who was nicknamed "Red," stayed on the Isthmus as a photographer until his retirement in 1937. At that time, his payment amounted to $292 per month.[71] He was the author of a book, a photo collection of battleships that passed through the Canal.[72]

It is interesting to compare Hallen's original salary to those of other Canal employees. Secretary Bishop received $10,000 per year, roughly five times of what Hallen was making. The photographer was paid about the same as a steam shovel craneman ($185 per month) or a locomotive engineer ($180) although considerably more than a cook ($100) or nurse ($95). This means that his work was regarded highly and on the same level with the rank-and-file engineers who built the Canal.

Hallen was given cameras already owned by the commission, among them a new Carlton 8 by 10 that seems to have been used for most of his photographs.[73] The negatives were gelatin dry plates, which, despite their superiority to earlier types of photochemical supplies, had to be handled very carefully. Hallen worked under extreme conditions: not only did he struggle with the obstacles of moving sensitive equipment through rough construction areas, but he also had to deal with tropical temperatures, rain, and mud.

A "systematic record" of all photos was called for,[74] and Goethals ordered that the plates be numbered, dated, and marked with "a description of the view."[75] He also set the conditions for the sale of photos to Canal employees, which officially started in 1908. Commission secretary Bishop received many requests from newspapers, and he indicated to Goethals that photographs were "a necessary part of the information which I am to supply."[76] Hallen's photos were regularly published in the annual reports of the commission informing members of Congress about the progress of the work.

In an article in the *Canal Record*, the official newsletter on the Isthmus, the author—most likely Bishop—explains that Hallen was compiling a "picture history" that would "furnish one of the most interesting and authentic accounts of the way in which the work was done."[77] He reports that negatives going back to the French project were part of this record, even though their quality and condition were often poor. Many prints and plates from the beginning of the American era in 1904 were missing, he notes, which implies that only under the Goethals regime did images become an integral part of the public relations effort.

Various sets of prints were made from Hallen's photographs. One, at certain times two, were designated for "Mr. Nichols' albums" (Nichols was the office engineer). It was decided that these sets would be stored in the fireproof vault of the administration building. One copy went to Goethals for his personal albums. Two, later three, were forwarded to the Washington office of the Canal Commission. One of these was to be made available for loaning; the third (from 1913 on) was designated for special access by newspapers and news services.[78] Employees of the Canal Commission were able to purchase Hallen's photos for 20 cents per print. The requests had to be made in writing and were subject to approval.[79] Demand for pictures increased significantly in the years prior to the opening of the Canal. Loan orders were "received from all sections of the country."[80] Since Hallen and his assistants lagged behind in their printing, the sale of photos to the Canal force and the general public had to be halted.[81] The previous order, or "circular" in commission speak, was revoked in 1911. From then on, photographs were only issued for "official use," Goethals wrote.[82] Another change of mind occurred shortly before the opening of the Canal in 1914, when the sale of photos for 20 cents each was again permitted but limited to fifty per order.[83]

Hallen's photographs were frequently turned into lantern slides for illustrated lectures on the Canal or use in schools. Chief Engineer Goethals was approached by prestigious institutions such as the National Geographic Society, Columbia University, and the Trans-Mississippi Commercial Congress and was asked to lecture on the progress of the work.[84] The New York State Education Department had a free loan collection of Canal lantern slides for schools and other educational organizations. Its Division of Visual Instruction published a list of images secured from the Canal Commission, which proved "an effective means of bringing to school children and the general public a clear notion of the position, size and nature of the canal and thus gives them the basis for a fuller understanding of its commercial and political significance."[85] During the final years of construction, the commission also received requests for motion pictures.[86]

The body of Panama Canal photographs held at the National Archives at College Park, Maryland, comprises about sixteen thousand images from the period of 1904 until 1939.[87] Some images dating back to the French construction effort are also included in this collection. The vast majority of photos were taken by Hallen, who remained official photographer for all but four years of the given time span, or under his supervision. Even though he made thousands of photos of one of the most important historic works of technology, no detailed study of these images exists. Ulrich Keller's analysis in the introduction to his collection of

Canal images is incomplete and occasionally inaccurate.[88] It is also tainted by judgments of Hallen by the same aesthetic standards that Krauss deconstructs in her work. Keller, for example, states that "his creative, artistic talents always remained modest."[89]

Hallen also served as a coordinator for the motion pictures made during the latter years of the Canal building. Although documentation is scant, it can be inferred that he was equipped with a very basic movie camera that then turned out to be defective.[90] Hallen stuck to his photography. A major effort to record the construction process on film was made by the Office of the Superintendent for the U.S. Capitol Building and Grounds in Washington, D.C. The photographer for this office, R. G. Searle, and an assistant arrived on the Isthmus in the spring of 1913 to produce an official motion picture—specifically to record the completion of excavation work in Culebra Cut.[91] However, the company that developed the film complained that many of Searle's images "are very unsteady, some of them are a little out of focus, and in others the panorama is not very clearly done," implying that the creator was an amateur. The firm then offered their own services for a follow-up visit to the Canal Zone.[92] Other motion pictures were made independently by film companies such as Selig and Edison, by other government agencies, and by the army. In general, the government did not consider motion pictures a major public relations medium in the way it had institutionalized the photography of the Canal building. Only in the final months of the construction effort, when cinematographic technology had become more advanced and was set to play a major role in the Panama-Pacific International Exposition, did the subject come up more frequently in the official communication.[93] Based on new shooting and editing techniques established by D. W. Griffith, film evolved into a powerful entertainment and propaganda tool. It is tempting to imagine how this medium would have changed the publicity and interpretations of the Panama Canal had its construction been carried out ten years later.

In conclusion, the records provide a good estimate of the role visual images played in the construction process. Although the engineers wanted to document the day-to-day progress of the work, Secretary Bishop and the commission office in Washington, D.C., used photographs to foster support for the project among politicians, the general public, and the canal laborers themselves. The photographs served as illustrations in schoolbooks and lectures to elite audiences. They were also routinely distributed to newspapers. Under Goethals, the Canal photography was handled in a systematic, managerial way. Cataloguing, distribution, and preservation were professionalized. Although Hallen was never referred to as an artist, he was paid well and his work regarded highly. Following Roosevelt's lead, the government was highly aware of the decisive role of images for the popular success of the Canal project.

Choosing the View: Perspectives and Panoramas

The publication and consumption of images taken on the Isthmus always took place in a context, as part of an annual report, a chapter in a book, a newspaper article, or a box of stereographs. What were the characteristics of these pictures? Which specific images were chosen to illustrate the building of the Panama Canal? And finally, how did the envisioning of the project express itself in the authors' texts and interpretations?

Although Hallen's photographs comprised the largest body of Canal images, they were not the only ones available to publishers and authors. News syndicate and freelance photographers and even tourists traveled to the Isthmus to record the building of the giant seaway. Perhaps the most striking aspect is the similarity of themes, motifs, and photographic technique. Hallen knew the Canal Zone better than anyone else, and he may have directed other photographers, who often came to the Isthmus on short assignments, to the same spots he chose for his own work. We can also assume that the government regulated access, especially to the construction areas, whether through official orders or "guided tours" by Hallen and other employees who led the guests to carefully selected locations.

The largest part of Hallen's images displayed the actual construction work, which is the main focus of this chapter. Other frequent topics were depictions of equipment left by the French, building exteriors (administration offices, hospitals, quarters) and interiors (mess-hall dining rooms, hotel lobbies), as well as city views. People played a lesser role in the overall body of photographs. However, they did figure prominently in views of everyday life in Panama. Hallen also took group portraits of commission officials, engineers, policemen, and the different ethnic groups of Canal workers. If we compare these themes to the photographs

President Roosevelt published in his message to Congress or to the vast collections of stereographs, we find the same categories.

The Panama authors made frequent use of Hallen's photos in their books. Frederic Haskin relied only on the images of the official photographer and used this fact to advertise his book. Of fourteen major publications examined for their use of photos, seven clearly printed reproductions of photos by Hallen, and in three cases this was most likely the case but could not be confirmed.[94] Two authors gave full credit to the source of their pictures, five only in some cases, and seven neglected the issue. With the exception of Haskin's work, Hallen (or the government) is never credited, supposedly because these images were distributed for free and without copyright. Among some of the existing credits are news agencies such as Underwood & Underwood and the American Press Association as well as individual photographers. Easy to identify are Hallen's photos of a West Indian worker spraying larvicide[95] and of a group of Italians laborers.[96] Also popular were before-and-after scenes that Hallen used to demonstrate the benign American influence on the country. A typical example shows a street in Panama City before and after paving, symbolizing the successful methods of sanitation (figs. 6 and 7).[97] This kind of arrangement of paired photographs was also used in depictions of colonial progress on the Philippines.[98]

A frequently reproduced photo showed the construction of the Gatun locks from a bird's-eye perspective (fig. 8). Panama author Ralph Avery explained that it was "unique in that it is the only one of its kind, having been taken from an especially constructed tower 157 feet high."[99] Hallen's photo, which is reprinted in Avery's book, is notable for its symmetry and division into three parts. The symmetry is created by the tower's position on the middle of three concrete driveways for railroad engines, which would later tow the ships through the locks. The foreground shows two basins of water extending to the margins of the photograph, whereas the middle part shows the land (a village can be seen on the left) through which the Canal and the locks were built. The background holds the horizon and a stretch of sky, where the onlooker can imagine Gatun Lake or even the Pacific Ocean. Avery's publisher cropped part of the bottom of Hallen's original photo so that total symmetry—both vertically and horizontally—was achieved. The photo of the locks became a symbol of the Canal itself, forcing its way through

Figure 6. A street in Panama City before . . . (National Archives, 185-G-150-A).

Figure 7. . . . and after paving, 1907 (National Archives, 185-G-150-B).

Figure 8. A view of Gatun locks from a temporary tower (National Archives 185-G-519-S34).

Figure 9. Culebra Cut in 1910 (National Archives, 185-G-3-J).

Figure 10. Culebra Cut in a lithograph by the artist Joseph Pennell (reprinted from *Joseph Pennell's Pictures of the Panama Canal* [Philadelphia: J.B. Lippincott, 1912], 15).

Figure 11. Shoshone Falls, Snake River, Idaho, by survey photographer Timothy O'Sullivan (National Archives, 106-WB-374).

the land, raising the power of American engineering and of a new, symmetric order literally to a higher level.

A similar perspective is applied in views of Culebra Cut, the eight-mile stretch of the Canal where most of the actual digging had to be done (fig. 9). Hallen's photo shows the bend of the cut and therefore gives up symmetry in order to underscore the differences in height that had to be overcome. In contrast to the photo of the Gatun locks, technology is not the main focus here—only a few pieces of equipment, among them a dark shovel on the right side, are scattered throughout the cut as signifiers. The artist Joseph Pennell produced an almost identical image that also includes a bend (toward the right instead of Hallen's left turn) and a steam shovel with an engine at the bottom (fig. 10).

Thus, the elevated perspective used in landscape painting and survey photography reappeared in the Canal photography. A comparison of Hallen's and Pennell's images of the cut to O'Sullivan's famous picture of Shoshone Falls suggests an influence (fig. 11). The landscapes are

Figure 12. The interior of an approach wall at Gatun locks, 1913 (National Archives, 185-G-519-013).

corresponding, even though the implication is different: whereas O'Sullivan's evocation of the sublime was based on the gigantic dimensions of nature, the miracle of Culebra Cut is the result of human engineering. In both cases, the perspective represents the appropriation of the land. In Hallen's interior view of an approach wall (fig. 12), another expression of the sublime, nature has disappeared completely. The image is highly reminiscent of the nave of a church, suggesting religious interpretations.

Hallen's images often explain the new in terms of the old. The concrete arches of the locks invoke the flying buttresses of gothic cathedrals.

Figure 13. A view of the new train station from Balboa Heights, 1915 (National Archives, 185-G-417-A7).

In the many images of the Culebra Cut, the visual language is the same as in earlier survey photography. Influenced by documentary photography, artists of the 1920s and 1930s showed men and machines in close-up shots, implying their exchangeability. Hallen rarely employs this perspective. Although his visual language may be aesthetically disappointing by today's standards—Keller writes of Hallen's "bland, unimaginative approach"[100]—it reinforced the message policymakers and authors were conveying to the general public. The building of the Panama Canal was a synecdochic project, and every image also represented the whole of the national effort, whether it showed Roosevelt atop the steam shovel or the digging of Culebra Cut.

The panoramic view became the preeminent perspective on the Canal Zone not only for the depiction of the construction work but also for the display of the social infrastructure on the Isthmus (fig. 13).

Going even further, it entered the *texts* of the Panama authors when they assumed a point of view for their descriptions of the Canal Zone. Canal Commission secretary Bishop put himself in the same position as Hallen's camera or Pennell's pencil when he described the work down in Culebra Cut:

> To stand at the southern end of the Cut, between the towering, majestic hills of the Great Divide, was an experience which few who had ever had it could easily forget. On either side were the grim, forbidding, perpendicular walls of rock, and in the steadily widening and deepening chasm between—the first man-made canyon of the world—a swarming mass of men and rushing railway trains, monster-like machines, all working with ceaseless activity, all animated seemingly by human intelligence, without confusion or conflict anywhere. Throughout the eight miles of the Cut the scene varied only in the setting. The rock walls gave place here and there to the ragged sloping banks of rock and earth left by the great slides, covering many acres and reaching far back into the hills, but the ceaseless human activity prevailed everywhere. Everybody knew what he was to do and was doing it, apparently without verbal orders and without getting in the way of anybody else. It was organization reduced to a science—the endless-chain system of activity in perfect operation.[101]

Technology is a major aspect of Bishop's observations, but the focus is not on the performance of synchronized motions or the production of identical goods (which, in the 1920s, dominated social criticism as well as industrial photography). Rather, he emphasizes the efficient organization of the work, the interlocking of complex actions, made possible by managerial rigor and commitment. His is a totalizing view on all of society, not a glance at details or an immersion into the scene before him. Only from a detached point of reference, without the noise of the actual work, without the impressions of exhausting, dangerous labor, could Bishop have made his generalizations as effectively as he did. The visual representation of the Canal building helped the Panama authors to direct their readers' view to the whole of the project, the "big picture." Some of the images were reminiscent of westward expansion, of the building of the nation by the pioneers. These photos showed the foreign country of Panama, but what the spectators saw, or were supposed to see and read about, was the construction of a new America.

"To Draw the Canal as It Is":
Joseph Pennell's Lithographs

Compared to the thousands of pictures Hallen took, the twenty-eight lithographs produced by the artist Joseph Pennell constitute only a small yet significant body of Canal images. Along with comments by their creator, the lithographs were published in a booklet that went through six printings within a single year.[102] Magazines and Panama authors such as Walter Pepperman included the reproductions in their pages. Museums and art galleries, among them the Uffizi Gallery in Florence, acquired full sets of Pennell's lithographs.[103]

At the time, illustrators fought to defend their work against the onslaught of photography, which was cheaper and regarded as more realistic.[104] In the age of the halftone process, these efforts were increasingly futile. And yet, one important advantage remained: in the first decades of the new century, lithographs were more readily viewed as art, and viewers were interested in what the creators had to say about their work. This may have been the reason why Pennell's images became the most popular rendering of the technical and social achievements in Panama.[105] In contrast to Hallen, who remained unknown to the general public, Pennell must be regarded as one of the Panama authors.

The artist showed little respect for photography. "I have been told I see these things through a temperament. I hope I do. I have no desire to pose as an artless artist or a pitiful photographer."[106] The outcome of his trip to the Isthmus was obvious from the beginning: "I went to the Panama Canal because I believed the greatest engineering work the world has ever seen would give me the greatest artistic inspiration of my life. I went because I believed that at the Canal I should see the Wonder of Work, the Picturesqueness of Labour, realized on the grandest scale."[107] Despite his proclamation that he wanted "to draw the Canal as it is,"[108] it is evident that Pennell came to Panama to put on paper (or rather, limestone) what he had expected to see.

The illustrator was born in Philadelphia and worked first as a clerk for a local coal and iron company.[109] Later he studied with landscape painter Thomas Eakins at the Philadelphia Academy of Fine Arts. He took several trips to Europe as a travel writer and artist, producing city views. During the 1880s he settled in London and did not move back to the United States until 1917, nine years before his death. Pennell's depiction of the evolving urban landscape of American cities often echoed European predecessors: New York's skyline, for instance, reminded him of the spires of Italian towers and churches.

Pennell also illustrated the journeys of the young William Dean Howells to Italy. On the Isthmus, he felt that he had rediscovered the spirit of the Mediterranean, "a mountainous country, showing deep valleys filled with mist, like snow fields, as I have often seen them from Montepulciano looking over Lake Thrasymene, in Italy."[110] He also compared the whole of the Canal Zone to "a perfect Japanese print."[111] In spite of the jungle, it conveyed charm and order to him. Observing the palm trees near Gatun, Pennell commented: "Through these wandered well-made roads, and on them were walking and driving well-made Americans."[112] At the Canal construction site, he discovered "splendid springing lines" carving "arches and buttresses" out of concrete, which resembled the structure of a cathedral or Roman aqueduct.[113] Pennell was sad to note that his artistic subjects would soon be buried under water and dirt.

In general, Pennell's twenty-eight images depict the same Canal scenes as those chosen by Hallen and other photographers to give an impression of life in the Canal Zone: construction sites (eighteen), village views (five), buildings (two), and nature (three). Each image was accompanied by a written comment. Pennell contrasted a "Native Village" with the mosquito-screened houses and palm-lined streets of the American settlements and showed a crane raising French machinery from the swamp. His depictions of the construction process are striking for their intense lines. The lithographs were created from the same perspectives and in the same contrast-rich optics that Hallen had selected for his photographs. By virtue of the drawing technique, however, Pennell was able to accentuate certain details and leave out others.

The image of the gates of Pedro Miguel (fig. 14), one of three related motifs, focuses on a relatively small number of distinguishable foreground elements such as tubes and pipes. The railroad track is the dominating feature and gives the image an impression of depth. Bishop, who read and corrected the proofs of Pennell's manuscript, commented: "It is impossible to convey in words anything approaching an adequate conception of the picture which the series of locks, with their massive, towering walls and their equipment of colossal gates, presents. It defies description, as it does the camera, even in its wonderful modern development, and can be portrayed only by the inspired pencil of a Pennell."[114]

Figure 14. The gates of Pedro Miguel locks in a lithograph by the artist Joseph Pennell (reprinted from *Joseph Pennell's Pictures*, 19).

The artist also showed an interest in human drama when he included a group of workers being lifted by a rope from the bottom of Gatun locks in two of his illustrations.[115] Pennell's perspectives reappeared in similar visual contexts, such as the cover of Bernhard Kellermann's novel, which envisioned the building of a transatlantic tunnel and was published one year later.

The Canal officials were pleased with Pennell's efforts. "Many of those engaged in the work had felt the sublimity and majesty of the spirit which hovered over it," Bishop explained, "but Pennell, with the clear and inspired vision of the great artist, felt it, saw it, and with his trained and sure pencil traced its outlines for all the world to see."[116] Chief Engineer Goethals put his own impressions in more simple phrases than his public relations secretary but nevertheless found great praise for the artist: "I cannot express in words the pleasure that these pictures give me, as they illustrate so clearly, forcibly and vividly the work, and portray actual conditions with a force which I did not think could be developed in a picture," he wrote in a letter to the illustrator.[117]

Pennell shared his aesthetic conservatism with Hallen and other photographers of the Canal. He found the inspiration for American culture and technology in European predecessors rather than in a departure from traditions. His emphasis on the resemblances of the Canal Zone to familiar landscapes may have been an attempt to make the foreign environment of the tropics appear less threatening. His descriptions also evoked Walt Whitman's "Song of the Exposition," in which the poet imagined American industry as the heir to European architecture and culture.[118]

Not all artists shared Pennell's sense of tradition. Asked to make suggestions for the decoration of the Canal locks, sculptor Daniel Chester French and landscape architect Frederick Olmsted Jr. turned down the request. The industrial design could not be improved, they argued.[119] In the same spirit, a contributor for *Current Literature* commenting on Pennell's lithographs wrote: "[T]here is a certain kind of art which grows out of the industrial and scientific processes. Some forms of art . . . follow utility and efficiency, and attain a beauty that is bound up in the perfect fulfillment of function."[120] This phrase recalls Bishop's description of the labor scene in Culebra Cut in which he aestheticized the social and technological process unfolding before his eyes. Going beyond Whitman, who viewed technology such as bridges and canals as a democratic network, Bishop likened the state, the public body, to a complex machine. Art, in this sense, means perfect order and total control.

People on the Isthmus:
Capturing Race and Ethnicity

As a mass medium, photography reflected the widespread racism of the era. Stereographs made fun of blacks and displayed them as amoral characters, providing cheap parlor entertainment for their viewers.[121] Photographs not only mirrored but also shaped and manifested hierarchies of identity. Coinciding with the proliferation of "scientific" methods and the expansion of the state, schematic visual records of people became widespread in areas as diverse as anthropology, the industrial workplace, and police and military investigation.[122]

The surveillance *gaze* of the dominant group and the repetitive, encyclopedic arrangement of images reduced people to objects. Oliver Wendell Holmes had predicted that photographic images would emphasize only the outer appearance of things and humans: "We have got the fruit of creation now, and need not trouble ourselves with the core."[123] From the final decades of the nineteenth century on, these surfaces were used to build potent stereotypes of their "core," and thus bodies became "the vehicles of gendered and racialized interior essences."[124] On the other hand, these attempts at inscription were resisted by the very objects of the photographic lens.

For the larger part, the visual archive of the Canal building consisted of engineering and construction images. The occasional group portraits of engineers and other gold-roll employees were traditional shots, usually taken in front of an administration building. Portraits of laborers and Natives were not very common. Nevertheless, the few depictions of workers did reflect the racial preconceptions of the Panama authors. In a letter to the photographer prior to his official employment, engineer F. B. Maltby criticized the scope and efficiency of Hallen's work: "On our first trip, I think the work took you at least two hours and all the pictures were made that we had any use for. Many of them are of no value whatever except for pictorial effect. A group of niggers shoveling dirt down a bank is of no value showing the progress of the work."[125]

In the *Canal Record*, the article on Hallen's photographs and the "The Picture History" of the project commented: "No class is neglected. West Indians, Europeans, and Americans alike, are represented."[126] Upon closer inspection, the photos tell a different story. The unskilled Caribbean laborers performed most of the physical work but received little credit for it. Other silver men from Italy and Spain, though smaller in number, were praised for their contributions. Most often, the Galegos

Figure 15. A group of Italian canal workers (National Archives, 185-G-244-B).

and Italians were depicted in groups, showing them either at work or posing proudly in working gear, with shovels and other utensils. Hallen's photo of a group of Italians, which was frequently printed, is an example (fig. 15). The laborers face the camera directly, hands on their hips—they obviously do not mind being photographed. Some authors paid little attention to which specific ethnicity was represented in a picture. In their book directed at schoolchildren, Alfred Hall and Clarence Chester, a history teacher and an explorer, described an often photographed group of digging Galegos as "A Gang of Italians." The text serves as an extended caption to interpret the photograph: "Of all the 'Silver Men,' the Spaniards and Italians are the best. They will do twice as much work per day as will the negroes, and receive much more pay. . . . The Spaniards are perhaps less likely to suffer from the climate and, therefore, accomplish more. They are small in size but muscular, willing to be taught, and anxious to be promoted to better positions as subforemen or foremen of their work. Where strength and intelligence are

Figure 16. A West Indian laborer applying mosquito spray in the Canal Zone, 1910 (National Archives, 185-G-640-A).

needed, these men can be depended upon. No amount of rainy weather can keep them from the work."[127]

The West Indian laborers were rarely depicted in groups. Most often they appear as single, isolated men not representing their ethnicity but performing a function. Examples are the popular picture of a worker spraying larvicide (fig. 16) and images that show laborers in front of or in the midst of a building site (fig. 17). In many of these photos, the worker is merely included to give the viewer an idea of the scale of the construction site. The human figure and, by implication, its effort are diminished in comparison to the spectacle of technology. In other cases, the West Indian worker is featured more prominently in the foreground in order to produce a depth effect when the image is viewed through a stereoscope. In figure 18, originally an Underwood & Underwood stereograph, we see a laborer standing in front of the Gatun emergency dam, an enormous steel structure. Because of his hat, the facial features can barely be distinguished—he becomes an anonymous worker.

The Caribbean laborers were also shown outside of the work environment. Their poor living conditions were a popular subject among the Panama authors. Photos such as one of a laborer's house (fig. 19), a

Figure 17 *(above)*. A West Indian laborer on top of the gates of Gatun locks, 1912 (National Archives, 185-G-518-Y14).

Figure 18 *(top right)*. A West Indian laborer in front of an emergency dam near Gatun locks (reprinted with permission from Keystone-Mast Collection, UCR/California Museum of Photography, University of California, Riverside).

Figure 19 *(bottom right)*. The house of a West Indian laborer in the Canal Zone (National Archives, 185-G-302-A3).

picture taken by Hallen, formally resemble the work of documentary photographers but lack the intent to draw attention to social ills. Photos of the West Indian quarters either served to illustrate their foreignness or their "primitiveness," especially when they were contrasted with the relaxed social life of the gold employees (fig. 20). A picture of West Indian homes in Colon taken during the French Canal project but resembling Hallen's scene bore the caption "Filth That Would Drive a Berkshire from His Sty."[128] Sadly, the comparison also demonstrates that little improvement in the West Indians' living conditions had been achieved under the American regime.

With regard to the Caribbean population, the authors noted a shyness in front of the camera. "As for the negroes, some live a life away off in the forests almost as wild as the Indians. They are not at all dangerous. Indeed, they are so timid as to be hard to photograph. Some must be caught and held before the camera."[129] The accompanying image,

Figure 20. The reading room at the Canal Commission club house in Culebra, 1909 (National Archives, 185-G-293-C).

originally an Underwood & Underwood stereograph, shows a naked boy escaping a man who tries to drag him in front of the camera. In Willis John Abbot's *The Panama Canal in Picture and Prose,* the caption for the same photo reads: "Trapping an Aborigine."[130] We must therefore assume that distinctions between the West Indian immigrants and the Native population ("Indians") seem to have been applied at random. The main purpose was to depict their allegedly strange behavior or inferiority. It is important to note that attempts to escape the camera and skeptical looks, as were sometimes depicted in images of West Indians, expressed resistance to photographic "capture" and transmitted this active struggle to the audience in the United States.[131]

Some Panama authors also used photographs to illustrate the "racial types" of the Canal Zone. Haskin, for instance, isolated people from group photos or larger scenes and displayed them as representatives of their ethnicity. The man in the striped shirt from Hallen's group portrait was used to exhibit "an Italian."[132] Abbot went further and included photos for chauvinistic and racist display. The image of a Panamanian girl carried the caption "The Soulful Eyes of the Tropics."[133] Other indigenous people, looking shyly into the camera, were shown nude or semi-nude. These pictures and the image of a Guaymi man ("Note the tattoed [*sic*] marking of face and the negroid lips") were credited to the *National Geographic Magazine.*[134]

The depictions of the natives and the West Indians were by no means unusual in the context of imperialism. A voyeuristic, racist display of other ethnicities was often accompanied by photos intended to demonstrate the potential for the "uplift" of the colonized. The racism of these photos could not be separated from its gendered nature. Anne McClintock writes that foreign people "were figured, among other things, as gender deviants, the embodiments of prehistoric promiscuity and excess, their evolutionary belatedness evidenced by their 'feminine' lack of history, reason and proper domestic arrangements."[135] It was precisely this impression that was achieved by the depiction of isolated, "passive" Caribbean workers as well as by the intimate, intruding portrayals of their families and appalling living conditions. And yet capture, the male act of subjection, was not always successful. The people displayed in these and other colonial photographs sometimes resisted the gaze imposed on them and made their own impressions on the viewer, if only by staring back or frowning.

Summing Up: The Super-Vision of Space

The evolution of photography was part of the technological tide that made the building of the Panama Canal possible. It was far from being the established medium of the second half of the twentieth century (before digital technology would lift photography to yet another level). The public demand for images was satisfied not only by printed illustrations in books, newspapers, and magazines but also by spin-off media such as stereographs and postcards. All of these helped to show and explain to middle-class Americans the rapid changes at the dawn of the modern age, from social and technological advances to the expansion of the American empire, bringing foreign people and places into the home. The Panama Canal was the first public project to encompass all of these transformations, and its visual appeal can hardly be overestimated.

In contrast to paintings and drawings, photographs pretended to depict "reality," when in fact they were highly constructed images. Their construction and interpretation were determined by the photographs' provenance, the instructions given to their producers, by the context in which they were published as well as by their modes of reception. Images were part of the discourse of making sense of the modern self and society and figured in this discourse as single photographs or series of photographs similar to paintings and other traditional visual media, but also as items in a larger photographic archive. Images created a new battlefield in the struggle over interpretations and meanings. Significantly, this new visual culture "does not depend on pictures themselves but the modern tendency to picture or visualize existence."[136]

Theodore Roosevelt had realized the potential of photography to promote the Panama Canal within the United States, and by posing on the steam shovel, he created one of its most powerful images. Under the Goethals regime, the government embarked on a complex project of documenting the construction of the waterway and establishing channels for the distribution of photographs. It exploited its position of power in the Canal Zone to control the production of images. The photographs subjected the Panamanian landscape to American control and excluded the West Indian laborers. They were shown only as isolated figures or in a racist, gendered context meant to prove their inferiority, and yet they sometimes managed to resist this labeling by staring back disapprovingly or escaping photographic capture.

Hallen's photographs and Pennell's illustrations, even O'Sullivan's earlier Darien images, often adopted familiar motifs from European architecture or American landscape in order to appropriate and contain the foreign and the unknown and evoke feelings of the sublime. Increasingly, as in Bishop's observations, the human impact on the environment, manifest in technology and efficient organization, rather than nature itself, was the object of such emotions. What American middle-class families were supposed to see through their stereoscopes were not the details of the work or the Panamanian landscape but a vision, a metaphor, of the future America.

The panoramic perspective used in survey photography was applied to the visualization of the Panama Canal, and the authors internalized this view in their writings. "I have tried to present to you the big facts of the Panama Canal and to give you a bird's eye glimpse of the wonderful engineering feat which is to unite the world's two greatest oceans," Hugh Weir wrote.[137] Not the close look but the totalizing frame of reference, the engineered view, became the preface for interpretation. Engineers, photographers, and Panama authors were, in Foucault's words, "managers of collective space,"[138] creating a "super-vision" for the audience at home.

4

Ideal Community

The Canal Zone as an American Utopia

The staggering dimensions of the Panama Canal and the wonder of the engineering work were the main factors that drew the public's attention to the Isthmus. The Panama authors devoted a large part of their descriptions and interpretations to the construction of the giant seaway. But they also directed their pens and camera lenses elsewhere — to the strange landscape of the tropics, the colorful people of Panama, and, most enthusiastically, the institutions and practices of the state that the U.S. government had erected in the Canal Zone. "Gradually there pieces itself together a splendid human story," the author John Foster Carr wrote in 1906, when the success of the Canal project was still in doubt to other observers. "You begin to understand that our Republic is doing something more on the Isthmus than the mere building of a canal. It is creating a state with all the machinery and equipment of our home civilization adopted to strange needs."[1]

The Panama Canal project shared many characteristics with colonial ventures and became part of the emerging American empire. And yet fundamental differences remained. Whereas the emphasis in other regions such as the Philippines was placed on the application of American social structures and institutions to the allegedly inferior society of the colony, the task in the Canal Zone required a new type of governance. In order to build the Canal and guarantee its operation in the decades to come, the commission had to set up a semi-permanent administration on the Isthmus that would organize the work and take care of the needs of its employees, including their social life. By definition, this mini-state had a single purpose: to produce and maintain the Canal. But most of the Panama authors chose to ignore this context of special circumstances and explained the society on the Isthmus as part of the larger project of constructing a new nation. The Canal Zone was turned into an American utopia.

These interpretations resulted in one of the most elaborate and enthusiastic endorsements of autocratic collectivism in American history. As the head of this regime, Chief Engineer Goethals reigned as a "benevolent despot." In this chapter, I examine how it was possible that the Panama authors discovered an ideal community in the midst of the tropics, a model society for the entire United States. Recalling familiar utopian blueprints—most importantly Edward Bellamy's bestselling novel *Looking Backward, 2000–1887,* which had intrigued middle-class readers for decades—the storytellers saw the Canal Zone in the light of the changes and challenges with which American society was confronted. It is puzzling that a close analysis of this utopian discourse has never been attempted in the historiography of the Panama Canal.

Suburban Fantasy: Civilization and Consumption on the Isthmus

From 1907 on, after the third Canal Commission had been installed, the positive comments on the Canal project were also applied to the government of the Canal Zone. The perception changed from the jungle of hell that the poet James Stanley Gilbert had described to a tropical Garden of Eden. With the deadly diseases of yellow fever and malaria under control and the necessary infrastructure installed, the regime on the Isthmus left a lasting impression on visitors. "This is the wilderness that for hundreds of miles has not known the foot of a white man, through which Uncle Sam's civilization-builders are stretching a twentieth-century wonderchain," wrote the journalist Hugh Weir.[2] To many observers, what had been achieved on the Isthmus concerned the entire nation: "The Canal Zone is the best governed section of the United States," the artist Joseph Pennell noted.[3]

In the spring of 1908, the first census of the Canal Zone was carried out. The total population amounted to 50,003, of which 24,963 were employed by the commission or the Panama Railroad Company. An estimated 11,091 additional employees lived in the cities of Colon and Panama outside the Canal Zone or had refused to declare their employment status to the censors. Of all inhabitants, 14,635 were described as white and 34,785 as black. The only other category, "yellow," was composed of 583 Chinese workers. By nationality, 6,937 were Americans and 11,411 were Panamanians (who were counted as blacks). The largest groups of West Indians were citizens of Jamaica (8,418), Barbados (6,483), and Martinique (2,397). The huge contingent of Spaniards (4,370) was the largest European population.[4] By the second census in the winter of 1912, the Canal Zone population had risen to 62,810. The number of commission and railroad employees had increased to 32,513, and the figure of American citizens reached 11,850.[5] The population figures for Colon (17,748) and Panama (35,368) were determined in a census carried out by the government of Panama in 1911.[6]

This data shows that within the multinational (and multiracial) community in the Canal Zone, the share of Americans reached levels between 14 and 19 percent—they were clearly a minority. At the same time, almost all of the U.S. citizens were white. The early census put the figure of "colored persons" from the United States at 73.[7] In their exultant depictions of the Canal society, the Panama authors usually referred exclusively to this white American population, which was predominately male—women made up 31 percent of the total in 1912.[8] The West Indians were considered only as a negative contrast, and since they held other citizenships (later mostly British), there appeared to be no need to incorporate them into the picture of the model American settlement.

What struck the writers most were the social benefits the American workers enjoyed. The engineers received better pay compared to similar positions in the United States and were entitled to forty-two days of paid vacation and thirty days of sick leave.[9] Medical treatment and housing, including furniture, were free. The employees resided in "comfortable if not elegant quarters . . . with a shower-bath handy and all janitor or chambermaid service free";[10] married men resided in spacious apartments. The meals in one of the commission's fifteen mess halls cost only 30 cents per day. One author described a typical dinner: "Mixed pickles; Rhode Island clam-chowder; lobster with mayonnaise; roast young turkey (stuffed) with cranberry sauce; French toast with fruit sauce; asparagus with melted butter; potatoes in cream; chocolate ice-cream; jelly cake, cheese, crackers; tea, cocoa, coffee. Can you surpass these meals—served in the Panamanian jungle—for the same figure at an American restaurant?"[11]

The Canal Zone enjoyed the low tariffs that Secretary of War Taft had arranged with the Panamanian government in 1904. Luxury goods were often cheaper than in the United States with its restrictive Dingley Tariff Bill dating back to 1897. Abbot mused that "it would take another article to relate the rhapsodies of the Zone women over the prices at which they can buy Boulton tableware, Irish linen, Swiss and Scandinavian delicatessen, and French products of all sorts."[12] The Canal Zone's commissary—nicknamed "Uncle Sam's giant department store"[13]—ran eighteen stores with constant low prices. The Canal workers made 90 percent of their shopping purchases at these locations.[14] After the establishment of the commissary, Panamanian salesmen were de facto priced out of competition. Conveniently, the commission employees were able to pay for purchases within the Canal Zone with checks directly charged to their salaries.[15] In addition to these supermarkets, the commission offered all kinds of services with state-of-the-art equipment: "government laundries, bakeries with automatic pie, cake, and breadmaking machines, electric-light plants, ice factories, plants for roasting coffee and freezing ice cream."[16]

Specific conditions accounted for these benefits. In contrast to the United States, there were no competitive markets in the Canal Zone in areas such as labor, infrastructure, and consumption. The luxurious conditions were financed by taxpayers' money. Strikingly, the authors often neglected these obvious explanations and described the Canal Zone as their desired version of progress. During the first decades of the twentieth century, the patterns of social interaction in the United States were changing rapidly. Industrialization had paved the way for mass products, national brands replaced locally produced goods, and department stores became the sites where people of all classes could explore the new abundance.[17] The storytellers helped their readers at home to become aware of this evolving culture of consumption. At the same time, they served as teachers who explained the Canal Zone as both a mirror of and a model for American society. Progress was not only technological—as expressed in the Canal engineering and the automated production facilities—but moral as well. "Conspicuous consumption," a phrase coined by Thorstein Veblen in his social satire *The Theory of the Leisure Class* (1899), was nowhere to be found on the Isthmus. The authors depicted the French as self-indulgent playboys. In their case, luxury was characterized by excess and therefore immoral. Under American leadership, unnecessary consumption (and financial profit) did not play any part: "The Commissary is there to supply a need, not to stimulate an artificial demand. . . . A metropolitan department store spends more for a full-page display advertisement in a daily paper than it costs the Commissary for publicity in a year," Arthur Bullard wrote.[18] The social infrastructure in the Canal Zone was viewed as a means to overcome the divisions within society. It resembled an idealized department store: all people would buy the same products in the same places and thus behave as classless consumers.[19]

The leisure activities in the Canal Zone were another popular topic. Some authors stressed the simple routine of the laborer's day: "Eat, sleep, and work is the monotonous round of the canal employee and most of them save money."[20] Others described the manifold opportunities to find distraction after work. The international committee of the Young Men's Christian Association (YMCA) had set up seven clubhouses and equipped them with libraries and games. A membership cost ten dollars per year. The main purpose was to keep the Canal workers away from the local saloons in Colon and Panama City.[21] The commission had ordered the majority of these 175 "watering holes" to be closed. The alcohol licenses for the remaining 33 saloons were revoked in 1913.[22] Clearly, the purpose of the leisure program was to ensure that the after-hours activities of the employees did not undermine their capacity for work. Discussion groups and athletic competitions abounded. In 1907 a representative of the American Federation of Women's Clubs came down to Panama and set up clubs for the Canal Zone women.[23] One author summed up the leisure activities for the year 1912: 96,000 spectators at film showings, 104,000 games of bowling, 15,000 participants in gymnastics courses, 420,000 books on loan at the libraries, and 491 theater performances.[24]

Not just the Panama authors expressed their surprise at the modern lifestyle in the Canal Zone, but also the thousands of tourists who visited Panama in the years prior to the opening of the Canal. Among them were members of Congress, journalists, and curious travelers who could afford the trip. The official guide William Baxter had spotted people from all walks of life: "In the first place there is no such thing as a race or class of tourists. These people come from all races and every conceivable social and economic class. . . . They are generally comfortable men and women of 50 or more, a few spinsters, and an occasional girl of near 20 years."[25] The tourists arrived mostly by ship from New York and New Orleans—the West Coast service had a bad reputation. The United Fruit Company, the Hamburg-American Line, the Panama Railroad Steamship Line, and the Royal Mail Steam Packet Company all maintained regular services to Colon.[26] After the convenient passage— "cooled by a system of artificial ventilation which assures a comfortable night's rest even in the warmest weather"[27]—the tourists resided at the Canal Zone's two fashionable hotels, the Tivoli in Ancon, completed just in time for President Roosevelt's visit in 1906, and the Washington Hotel in Colon, built in 1912. Both hotels became centers of the social life in the Canal Zone.

With their personal needs taken care of, the tourists experienced the Canal Zone from an artificial perspective that had become increasingly accepted and popular by the end of the nineteenth century. In the familiar environment of hotels and guided tours, the encounter with the strange foreign country lost its threatening potential.[28] No immersion into the alien culture was required. The visitor's stay was often limited to a few days. Many Panama authors had a similar point of reference and little desire (or chance) to broaden their horizons. The most frequent type of critical comments refers to situations in which the accustomed or expected comfort could not be provided. A woman who traveled with her husband to the Isthmus found that her visit to one of the commissary's stores had been fruitless: "But we could not buy anything, no, not even though we had carried a million dollars in our pockets, because we did not work for the government. And I wanted a new veil, too."[29]

To the Panama authors, the Canal Zone displayed many elements of utopian landscapes and societies. The previous image of hell had turned into its complete opposite, and it now fitted the traditional description of the American tropics with references to El Dorado and other mystical places.[30] Abbot reinvented Panama as a tropical island: "Despite its isthmian character, the Canal Zone, Uncle Sam's most southerly outpost, may be called an island, for the traveler's purpose." His description of the isle, "walled in by the tangled jungle where vegetation grows so rank and lush that animal life is stunted and beaten in the struggle for existence by the towering palms, clustering farms and creeping vines,"[31] mirrored the material riches of life in the Canal Zone instead of the darkness and danger the tropics had represented at the outset of the project.

The American settlement was an assortment of chosen people, the authors argued. The Canal workers were "picked men," wrote Scott. "They must be in sound condition when employed and usually in the prime of life."[32] The beauty and youth of the zone's inhabitants was a frequent theme. When Charles Francis Adams observed a group of young girls at the Culebra train station, he mused: "A more healthy, well-to-do and companionable group of children could not under similar circumstances have been met at any station within twenty miles of Boston."[33] Other commentators rejoiced at the sight of the dancing crowd every Saturday night at the Hotel Tivoli: "The youth and beauty of the American contingent turned out in force at these functions."[34]

On the Isthmus, nature and culture merged into a suburban fantasy. Like the photographers, Bullard chose an elevated perspective, a hill near Paraiso, to describe the panorama that was unfolding below him: "On the sides of the hills you see villages—clusters of homes, well-kept lawns where all that is beautiful in the jungle has been separated from what is noxious and brought under cultivation: noble groups of palms, red and yellow and green shrubbery, flaming bushes of hibiscus; you see mothers in crisp white dresses playing with their babies; and if it chances to be the right hour, you will see a rout of children, as husky youngsters as you could find in East Orange, tumble out of the school house."[35]

The harmony of these random visual impressions was then projected into the social analysis of the Canal Zone. Utopia was placed in a new suburban realm combining the best of two worlds: "It has the intimacy, the everybody-knowing-everybody-ness of a country village and the tastes, the culture, the books and the gowns of the city—after they have had a few weeks to ripen," Bullard continued.[36] The malaise of the industrialized cities and the narrowness of the American countryside lifestyle were nowhere to be found in Panama. An atmosphere of mutual respect ensued, as fellow writer Ralph Emmett Avery noted: "The bringing together of people from every part of the United States, and the consequent interchange of ideas has given birth to a spirit of tolerance, of a broadening of the mind, and has led to the abandonment in a large measure of narrow-minded prejudices embodied in the selfish thought that 'My way is right, yours is bound to be wrong', a rut that people in small communities in the States are so prone to fall into."[37] In the minds of the Panama authors, the "big job" that everyone was working on had given birth to a perfect society: "Indeed, it was an ideal

community, from the pragmatist's viewpoint, that had its being in the
Canal Zone, for there was no privileged class, no idlers; every adult had
a part in the great project that had brought 35,000 men with their families together there."[38]

The storytellers painted the picture of a harmonious life, a fountain
of youth and a Garden of Eden on the Isthmus. In Whitman's "Passage
to India," the foreign land was no longer a place based on reality but a
utopian form of consciousness. In a similar way, the Panama Canal,
now the Passage to India come true, had paved the way for a fictional society. In addition to the benefits of the American pioneer community—
Turner's vision of the West—it had the cosmopolitan and technological
flavor of the modern era. It was a fictional world because it had unfolded
in the minds of the Panama authors. Even though Pepperman had included the West Indian workers in his figure of 35,000, it was clear that
the writers had drawn their enthusiastic conclusions from looking only
at the way of life of the American population. They did acknowledge
the social inequality and racial segregation that had been manifest in
the Canal Zone since the very first day of the work. But there was no
room in utopia for a revision of the American success story.

For the silver men, conditions in the Canal Zone were everything
but ideal. They lived either in barracks with bunk beds for 72 men
erected during the French era or in simple, self-built huts, often located
in the slums of Colon and Panama City. Only in 1913 did the Canal
Commission decide to build permanent homes for them.[39] The Panama
authors either thought of the inequality as natural (the housing facilities
of West Indians and Americans "of course differ radically," wrote Marshall[40]), or they laughed at the desperate attempts of the West Indians
to escape their assigned quarters: "As to the negro, his quarters were
barracks, with bunks that probably reminded him too much of the
bunks in a prison cell. So it happened that although here he could
have free quarters he greatly preferred a thatched hut in the 'bush.' . . .
Then he could invite his family from Jamaica or Barbados to the Canal
Zone, which invitation was promptly accepted by wives, children, dogs,
chickens, and all."[41]

Some authors criticized granting the Caribbean laborers any rights
at all (one journalist thought of the Canal Zone as a "nigger heaven"[42])
or made fun of the fact that they could actually take their pay home:
"The West Indian negro who saved was able to go back home and
become a sort of Rockefeller among his compatriots."[43] The census
enumerator Harry Franck, one of the few authors who had actually

lived in the Canal Zone, did not agree with the segregation of silver-
and gold-roll employees, yet he shared his colleagues' prejudices:
"There can be no question of the astounding stupidity of the West Indian rank and file."[44] Their alleged laziness and stupidity were qualities
that almost all authors mentioned. "They are very peaceable and law-
abiding fellows, but exceedingly lazy and unbelievably stupid," Farnham Bishop wrote.[45] Typically, these attributes were understood as the
opposite of manly behavior and contrasted with those of the American
workers: "They are thin, slow-moving, impassive, often solemn. There is
no glow in the dead yellows and browns of their flesh. But when you
look at our engineers, mechanics, and foremen, you see full-blooded
health shining in their faces."[46] These descriptions of the West Indians
were in line with the way in which African Americans were often portrayed as "unsexed primitives."[47]

The writers also reflected on the reasons for the presumed differences
between the ethnic groups and the influence that living in the tropics had
on human beings. Once again, they employed the jungle as a trope that
could connote whatever meaning suited them best. Some argued that
the tropical abundance made work seem unnecessary: "The Jamaican
negro is a natural loafer. Of course he works when he must, but betwixt
the mild climate, the kindly fruits of the earth and the industry of his
wife or wives, that dire necessity is seldom forced upon him."[48] Others
viewed the Caribbean laborers as barbarians who still had to climb the
ladder of civilization. Contrasting women doing laundry by the river
with a shiny new train platform with a newsstand across from them,
Weir noted: "Within a hundred feet are the two halves of my picture—
civilization and savagery, jungle and clearing, the primitive customs of
the black man and the modern progress of the white man."[49] In this
case, the jungle was associated with darkness and wilderness.[50]

The observations the authors had made with regard to the culture of
consumption in the Canal Zone were also tested against the behavior of
the West Indians. Bullard saw in the elaborate dresses of the Caribbean
women, "worn solely as a decoration, and not at all from a sense of
modesty," an example for Veblen's "Theory of Conspicuous Waste," as
he called it.[51] Similar to the depiction of the French, the supposedly immoral consequences of consumption were projected onto a foreign
group.

The bland racism of the Panama authors was in line with the descriptions of colonized people in other parts of the world, and its emphasis on

the "passivity" of the West Indians—which found its equivalent in the visual images of the construction—underscores the gendered nature of this discourse. The colonial subjects in the Philippines, as Michael Adas notes, were described as childlike and lacking initiative, unable to solve their own problems. This depiction served to contrast the alleged Native mentality with the virile discipline of U.S. engineers.[52] If the Filipinos adopted American ways, they would fare better. If civilizing efforts for "uplift" failed, however, their lack of success could be blamed on the Natives' effeminacy. As opposed to the "proper" colonial project in the Far East, building the Panama Canal had little to do with improving the life of West Indians, but sometimes a similar note was struck in the comments of the Panama authors. In a pensive mood, Canal Commission secretary Bishop contemplated the effects that their participation in the Canal project may have had on the West Indian workers: "It came about, therefore, that by introducing discontent into the daily life of the West Indian, the American canal builders made him a better laborer and a more useful member of society. White dwellers in those West Indian islands to which natives have returned after working on the canal say that they exhibit a marked increase in capacity. Whether the improvement will be permanent or not remains to be seen, but that it was made through injection into their lives of new and unsatisfied desires, with the consequent discontent, is the unquestionable fact upon which the sociologists of the world may concentrate their minds."[53]

Another group that was commented on but that certainly was not the focus of the authors' attention were the American women in the Canal Zone. Their main task, the male writers noted, was to play "the cheery wife at the front door."[54] But not every woman was satisfied with this limited role: "At Panama her housekeeping duties were lightened by the excellence of the commissary system, so that they were not enough to keep her mind occupied. She became homesick and hysterical."[55] This reasoning also indicates the changing gender roles in the new consumer society, as time-saving technology allowed women to spend more time on occupations other than housework. In general, women played a minor part in the writers' depiction of the Canal Zone. "For the married man the place is a paradise—if the wife likes it," Bullard wrote. "But after all it is a man's job and a man's community."[56]

The Panama authors described the Canal society as a predominantly white, male utopian community in which production—building of the Canal—and consumption were linked through practices representing both technological and moral progress. The West Indian majority was described in blatantly racist, derogatory comments in order to emphasize the superiority of the American model state. The writers depicted the Canal Zone as a suburban fantasy in which social innovations could be reconciled with the communal vision of the pre-industrial era—if only for a privileged minority. Applying tropical metaphors at random, the storytellers rarely stepped outside of the tourist perspective, indicating that the subjects of their interpretations were in fact the challenges facing American society at home rather than the actual conditions on the Isthmus.

Looking Backward: Edward Bellamy's Legacy in Panama

Many Panama authors went beyond the vivid, yet often disconnected observations of a utopian lifestyle in the Canal Zone and began to analyze the underlying structure of the state on the Isthmus and to compare it to earlier as well as contemporary economic and political patterns. The centralized Canal Zone government had little in common with the laissez-faire approach toward business practiced at home. Until then, the United States had been the epitome of a competitive, liberal economy. German tourists, for instance, pointed out the low degree of government regulation compared to their own country.[57] When the Americans took over the Canal project, hopes were high among Panamanian salesmen that the cooperative system with fixed prices, which had been introduced by the (then private) railroad company in 1894, would soon be terminated: "Under the United States flag the merchants of the Isthmus hoped for better things. They immediately petitioned Washington to abolish this iniquitous assault on private profits. They said they relied on the long established principles of our Government and its known abhorrence to stifling individual initiative."[58] As it turned out, the merchants had to abandon their hopes. The Canal Zone government moved even further away from free commerce and toward a centralized, autocratic rule. Puzzled at first by a political and economic system they had never encountered before, the Panama authors soon began to praise its efficiency and alleged justice.

The Canal Zone's administration was "surely contrary to the industrial dogma we have always been taught," noticed Bullard.[59] The state-run cooperative had removed the middlemen who sold and resold goods

for profit. Lower prices and "the elimination of graft" were the consequences.[60] The term "graft" referred to the controversial profit margins
these intermediate traders secured for themselves. It was widely used by
Roosevelt and other Progressives to denote corruption, and it had also
been applied to the French in Panama. The cheap, and in many cases
free, supply of housing, food, and health services as well as a state-
owned railroad, unheard of in American history, seemed to resemble
collectivist economic systems. It was time to part with previous beliefs,
one writer argued. The success of the state on the Isthmus "has shown
the absurdity of the ancient superstition that organized society, the
state, cannot attend to the needs of the people as economically and with
as efficient service as can an individual or a corporation."[61]

The authors used different expressions to define the administrative
system in the Canal Zone: "co-operative commonwealth,"[62] "military
paternalism,"[63] or even "modified socialism."[64] The use of the latter
term was highly problematic since it implied an ideological position on
the far left. The writers therefore used all kinds of tricks to circumvent
the "S-word." Some tried to explain the system as a result of special
circumstances ("the necessities of the situation,"[65] "a process of evolution"[66]) rather than of a calculated political decision. Others quoted
anonymous people who described the Canal Zone as a socialist state—
so they would not be identified with the opinion.[67] The *Atlantic Monthly*
printed a fictitious conversation, in which seven men with professional
and intellectual backgrounds—typical readers of the magazine—took
part. One of them, who was nicknamed "the Pest" and who, according
to the narrator, was "an infernal nuisance,"[68] made the other men guess
which socialist state he had visited recently, without giving away at first
that he was referring to the Canal Zone:

> "I am sticking to the facts," insisted the Pest. "It's all perfectly true,
> it's all happening every day, only you fellows are too busy theorizing
> about the labels on things to scrutinize their contents. Consequently,
> your ignorance of this state is wholly natural, because the founders of it
> are wholly unconscious that it is a Socialist state, and have never adver-
> tised it as such. In fact, if they were ever to learn that their governmen-
> tal activities were described in such terms, they would be horrified be-
> yond belief."
>
> "Do you mean to say," demanded the Real Socialist excitedly, "that
> this state has simply made up its own Socialism spontaneously, as it has
> gone along?"

"Precisely," said the Pest. "Paying no royalties whatever to Carl Marx or subsequent patentees."[69]

Although Progressive politicians such as Theodore Roosevelt expressed resentment toward big business and favored the regulation of industry, they avoided the terminology of organized socialists and communists and even used such labels to discredit other public figures. In a letter to his friend Lyman Abbott, publisher of the *Outlook*, Roosevelt directed his anger against Eugene V. Debs, the leader of the Socialist Party of America, and the social reformer Jane Addams: "Under the chin of disguise of standing for a movement of social reform, these different pamphlets and others like them are largely mere pieces of pornographic literature; just as Debs' paper and speeches are largely mere pieces of literature of criminal violence. One of these parlor socialists the other day, in addressing a girls' college, told them that 'motherhood was the curse of women.' Miss Jane Addams, in her recent book, shows lamentably by her own utterances the effects of belief in the socialism which bases itself upon Tolstoi (himself a sexual degenerate, whose *Kreuzer Sonata* is a fit supplement to his 'My Religion,' for erotic perversion very frequently goes hand in hand with a wild and fantastic mysticism)."[70]

Roosevelt's fierce conclusion was that "the Debs type of socialist points the way to national ruin as surely as any swindling financier or corrupt politician."[71] Debs, a former labor leader and veteran of the Pullman strike in 1894, had founded the Socialist Party of America in 1901. It was the most successful socialist movement in American history. In the election of 1912, he received more than 900,000 votes. Comparisons of the Canal Zone to a socialist state certainly spurred interest in the books and articles of the Panama authors, but with a broad middle-class audience in mind, they had to be careful not to endorse this classification themselves.

Some arguments against the label "socialism" in Panama shed light on the actual nature of the state on the Isthmus. "At no point does the canal project affect a complete economic operation. Money is being spent but it is not being made. The work is being done without regard to its ever paying," wrote Scott. The Canal Zone's wealth was based on "the gratuitous fruit of taxation."[72] The socialist cited by fellow writer Bullard made the same point: "Socialism will have to be self-supporting."[73] But for the most part, the objection to the terminology did not mean that the Panama authors would abandon their vision of an ideal state on the Isthmus. Naive as this may seem, there was a reason for their persistence: the Canal Zone brought back to mind a powerful literary utopia invented by a social critic from Massachusetts more than twenty years earlier: Edward Bellamy's America of the year 2000. The author's utopian novel *Looking Backward: 2000–1887* was published in 1888 and quickly became one of the bestselling books in American literary history.[74] In *Looking Backward*, Bellamy imagined a centralized and rationalized state offering social equality. It was the most influential of a few hundred utopian novels that were published during the final decades of the nineteenth century and until World War I.

"The dream of the late Edward Bellamy is given actuality in the Zone where we find a great central authority, buying everything imaginable in all the markets of the world, at the moment when prices are lowest—an authority big enough to snap its fingers at any trust—and selling again without profit to the ultimate consumers," Abbot noted.[75] His observation was one of two explicit references by the Panama authors to the novel. Fellow writer Franck commented on the nature of the commissary system: "Besides the hotel there is the P.R.R. commissary, the government department stores. It is likewise laundry, bakery, ice-factory; it makes ice-cream, roasts coffee, sends out refrigerator-cars and a morning supply train to bring your orders right to your door—oh, yes, it strongly resembles what Bellamy dreamed years ago."[76] Going beyond these direct tributes to the author, the characteristics of Bellamy's utopia were woven into the fabric of the authors' interpretations of the state on the Isthmus. Judging from Herbert and Mary Knapp's description in their book *Red, White, and Blue Paradise*, dropping Bellamy's name in the Canal Zone was commonplace even decades after the completion of the waterway.[77] Surprisingly, these comments made by two expatriate high school teachers putting together their post–World War II memories of the zone have so far remained the only reference to Bellamy's novel in the secondary literature on the building of the Panama Canal.

Looking Backward is the report of a wealthy young man, Julian West, who falls asleep in the basement of his Boston apartment in 1887 and wakes up 113 years later. Digging in his garden, the retired physician Dr. Leete discovers the visitor from the past. West had survived a night fire, which destroyed the rest of the house and prompted his friends to pronounce him dead, but which left the isolated room intact. The survivor describes the first week of his life in a future Boston, his guides being Dr. Leete with his wife and daughter Edith. West falls in love with the young woman—who, as it turns out, is the great-granddaughter of his former girlfriend. The conversations between West and Dr. Leete dominate the narration. The vision of a new world without classes and conflicts, which Bellamy elaborates in these debates, is the opposite of his perception of nineteenth-century America, strangled by social and economic crises.

In Bellamy's utopia, "the present system of organized production and distribution" (178) under the auspices of the government has eliminated private companies as well as the profit-seeking middlemen. Imposing supermarkets, "great distributing establishments in each ward of the city" (81), charge purchases through a cash-free credit system— similar to the commissary's stores in the Canal Zone. In every neighborhood, there is a "general dining house" (107) where the inhabitants have at least one meal per day. Vast differences in income and wealth no longer exist, as Dr. Leete explains: "We might, indeed, have much larger incomes, individually, if we chose so to use the surplus of the product, but we prefer to expend it upon public works and pleasures in which all share, upon public halls and buildings, art galleries, bridges, statuary, means of transit, and the conveniences of our cities, great musical and theatrical exhibitions, and in providing on a vast scale for the recreations of our young people" (165).

The nationalized factories employ a workforce called the "industrial army." Every citizen is required to serve in this army from the age of twenty-one to forty-five—even women, but they are separated and work "under [an] exclusively feminine regime" (174). The working population has no voting rights. "That would be perilous to its discipline," explains Dr. Leete (135). The state is ruled by an administrative elite, "a body of electors so ideally adapted to their office, as regards absolute impartiality, knowledge of the special qualifications and record of candidates, solicitude for the best result, and complete absence of self-interest" (134). These rulers are recruited from the retired workforce, based on their previous merits and qualifications. From their midst the most important positions in the country, including the presidency, are filled. Bureaucratization and the professionalization of expert groups such as doctors—a development that was in its early stages in 1888—characterize the efficient social organization in the state of the future.

Bellamy's society was based neither on a denial of technological development—like many agrarian utopias in the twentieth century— nor on a full embrace of industrialism. He was appalled by the transformation of American cities and the social tensions of his age. In *Looking*

Backward, he imagined a cleansed version of this deficient status quo, coming surprisingly close to our own post-industrial consumer society. Bellamy predicted the invention of many of today's key technologies such as television ("electroscope"), radio, and credit cards.[78] As the historian James Gilbert points out, most of his inventions focused on the delivery and distribution of products—including abstract goods such as information and music—and equal access to them. "In general, they have little to do with machinery, industry, mechanical invention, or production, all of which appear to be magically transformed, humming quietly off stage."[79] In many ways, the depictions of the Canal society matched this pattern. While the Panama authors celebrated the latest technological equipment in the Canal Zone's bakeries and ice plants, special emphasis was put on the cheap and easy use of these and other facilities, making it possible, as Franck notes, "to bring your orders right to your door." The zone's suburban layout differed drastically from the dense urban mix of factories and apartment buildings in the United States. Production was concentrated in the Canal building, uniting the "industrial army" in a shared project of national dimensions.

Although the actual military has been discontinued and even prisons have been abandoned by the year 2000, Bellamy uses the army as his most effective symbol of centralization. Having returned to his old life in a dream, narrator West is taken aback by the chaos on the streets, until he spots a military parade: "It was the first sight in that dreary day which had inspired me with any other emotions than wondering pity and amazement. Here at last were order and reason, an exhibition of what intelligent cooperation can accomplish. The people who stood looking on with kindling faces—could it be that the sight had for them no more than but a spectacular interest? Could they fail to see that it was their perfect concert of action, their organization under one control, which made these men the tremendous engine they were, able to vanquish a mob ten times as numerous? Seeing this so plainly, could they fail to compare the scientific manner in which the nation went to war with the unscientific manner in which it went to work? Would they not query since what time the killing of men had been a task so much more important than feeding and clothing them, that a trained army should be deemed alone adequate to the former, while the latter was left to a mob?"[80]

Comparing this quote to Canal Commission secretary Bishop's observations in Culebra Cut, we find striking similarities. While Bellamy mentioned the army's "perfect concert of action," Bishop spoke of an organization "in perfect operation." Both mention the scientific manner of the procedure they are describing, "all animated seemingly by human intelligence, without confusion or conflict anywhere."[81] It was precisely the confusion and conflict they were exposed to in the rapidly changing industrial society around them that upset both the utopian author and the public relations manager. On the Isthmus, the "army" was employed not to fight a war but to build a canal and a new society. Military metaphors had been useful to promote Roosevelt's program for the reconstruction of a manly and moral America. As modern society was getting more and more complex, they were also suitable to describe the trend toward rational, efficient, and centralized forms of organization.

Looking Backward became part of a literary phenomenon, a "tremendous outpouring of utopian literature" in the United States.[82] More than two hundred works were published during the following years and decades, among them Mark Twain's *A Connecticut Yankee in King Arthur's Court* (1889), Ignatius Donnelly's *Caesar's Column* (1890), and William Dean Howells's *A Traveller from Altruria* (1894), none of which came close to the popularity of Bellamy's novel. There were many mutual influences—Laurence Gronlund's *The Cooperative Commonwealth* (1884) is seen as a little-known model for *Looking Backward*[83]—and differences, meaning that some works fell into the category of dystopian visions. But in contrast to twentieth-century works such as Aldous Huxley's *Brave New World* and George Orwell's *1984*, in which the evolution of technology and the expansive role of the state results in the abuse of power and new social problems, most authors were convinced that the current crises could be overcome once and for all by replacing rampant individualism with a more altruistic public spirit. In a study of twenty-five "technological" utopias such as Bellamy's, Howard Segal concludes that "alterations were really conservative explorations from trends in existing society."[84] Utopia was at hand—it was the consequence of a Lamarckian transformation similar to the metamorphosis Roosevelt had propagated on a personal and national level. "I believe that when you make a fuller study of our people you will find in them not only a physical, but a mental and moral improvement," Dr. Leete tells his guest (180).

Many utopian writers extended this process to foreign policy. In their works, Susan Matarese identified "manifestations of a belief in America's greatness and its potential to play a singular role in world affairs."[85] References to European nations were often made only to set the future America apart from socialist models. Invoking Whitman's "Song of the

Exposition," the writers believed that the United States was destined to fulfill European aspirations and dreams in a new century.[86] Similar to the interpretations of the Panama Canal, these nationalist visions coincided with a call for international brotherhood. While on a trip to Europe as a young man, Bellamy had visited the cramped working-class neighborhoods (Gängeviertel) in German cities.[87] In his vision of the year 2000, European nations have undergone the same transformation as the United States, allowing for friendly relationships and free trade. "If I were to give you, in one sentence, a key to what may seem the mysteries of our civilization as compared with that of your age, I should say that it is the fact that the solidarity of the race and the brotherhood of man, which to you were but fine phrases, are, to our thinking and feeling, ties as real and as vital as physical fraternity," Dr. Leete explains.[88] To Bellamy, "civilization" and "race" were identical: African Americans play a negligible role in his novel.

Looking Backward was translated into many languages, even Chinese and Japanese.[89] Although the numbers vary, it seems certain that the number of copies sold within a year after publication was a six-digit figure.[90] By 1900 the popularity of *Looking Backward* was surpassed only by *Uncle Tom's Cabin*. The book even inspired a political movement, the Nationalist or Bellamy Clubs, whose members discussed ways to implement Bellamy's ideas. In 1890 there were 158 of these clubs nationwide, 65 of them in California and 16 in New York City. As a forum for discussion, Bellamy published a magazine, the *New Nation*, from 1891 until 1894. The movement never agreed on a coherent political platform, and by the mid-1890s most of the clubs were dissolved again. Their members turned to other reformist groups such as the Populists, or to new Social Democratic initiatives.[91] Bellamy also encountered criticism, especially from the Left, and responded in 1897 with a follow-up novel, *Equality*. The sequel was supposed to refute accusations that Bellamy's society was a tyranny in disguise. Compared to its predecessor, *Equality* raised little interest, as public attention turned toward the Spanish-American War.[92]

The mutual distrust of the workers' movement and the middle-class utopian authors resurfaced in Bellamy's writings. Although Socialist Party leader Debs respected Bellamy, he denounced the Nationalist movement as "Yankee Doodleisms of the Boston savants."[93] Bellamy, on the other hand, wrote to his literary mentor William Dean Howells: "I may seem to outsocialize the socialists, yet the word socialist is one I could never stomach. It smells to the average American of petroleum, suggests the red flag and all manners of sexual novelties, and an abusive tone about God and religion."[94] In *Looking Backward,* hindsight from the year 2000 showed that the anarchists were traitors.[95] For Bellamy, whose cousin Francis coined the words for the Pledge of Allegiance in 1892, a revolution resulting in rule by the masses was not desirable. In his novel, power is held by an elite, not the industrial army. "The refined drawing-room culture of sentiment and confidence, individualism and ideality, reappears in the noncapitalist environment of the year 2000," writes literary historian George Cotkin. "Bellamy's critique of capitalism is thus contained by its inability to extend to the area of cultural control. Authority, rather than freedom, is the key ingredient in the utopian experiment in *Looking Backward.*"[96]

The oppressive atmosphere of the new society was also a consequence of the author's personal beliefs and anxieties. Bellamy, who came from a family of Baptist ministers, was strongly influenced by Protestant ethics suggesting that rationalization and self-control went hand in hand. Fearing the subversive impact of excessive consumption, many of the utopian authors (and the Progressive reformers) imagined a smoke-free, non-alcoholic America, populated by traditional families who shared a Christian vocabulary.[97] Bellamy's inhibitions were also rooted in a frustrated search for self-expression. A trained lawyer, he was unwilling to practice in this profession or others and felt isolated from society. Longing for solidarity, he turned his society of the future into "a vague dream of unity and harmony—a utopia where the diversity that spawns distrust has vanished."[98] Even though consumption and leisure time play an important role, he never evokes feelings of sensuality or spontaneity. One Panama author conveyed a similar impression of life in the Canal Zone: "The same paternalistic commissary that reduced the cost of living and made housekeeping so easy, also tended with socialistic frankness to bring everybody to a dead level. It was useless to attempt any of the little deceits that make life so interesting at home."[99]

What do the parallels between Bellamy's vision and the state on the Isthmus indicate? The popularity of *Looking Backward* implies that the Panama authors and their audience at home were familiar with his ideas. More than twenty years after the publication of the novel, none of the radical changes he suggested had been realized. Nevertheless, the transformative character of Bellamy's and other contemporary utopias made them persist. Their new world was not to be found somewhere else—up in heaven, for example, as the Puritans were convinced, or out in an imaginary West, as Jefferson and others had believed. Instead, it was right in front of the eyes of everyone who dared to dream. The ideal future was an almost natural result of progress, eliminating the defects of the present as well as the fear of disruptive social change in the mind of the dreamer. This search for order left little room for deviation or dynamic change, so the new world evoked a "sense of completion, even stasis."[100]

The Canal Zone resembled these literary utopias: it was a laboratory, an enclosed space, where visitors and interpreters could discover the future America just like Julian West had suddenly opened his eyes on the future Boston. Recognizing social structures that had been described before in fiction, the Panama authors viewed the Canal Zone as an experimental yet familiar landscape, where the necessities of the Canal building had been turned into white, middle-class fantasies. The Canal Zone was seen as a showcase and a further step toward the realization of a cooperative society: "This is no idle dream, and within five years, yes, within three years, it will begin to be felt in the United States," Abbot quoted an enthusiastic businessman on the Isthmus as announcing.[101] Other authors came to the same conclusion. "This is the lesson of Panama," Bullard wrote. "The facts of the case force us to revise our old judgment. 'Collective activity'—this new force which we are developing with such amazing success in the tropics, which we, Americans, have carried further than any other nation—is worth considering as a means of solving our problems at home."[102]

Canals, bridges, railroads, and other public works in the United States as well as Europe often prompted utopian visions. In some cases, the view that these projects would unite the people of a community or nation spiritually and, by extension, promote universal peace was linked to specific social or philosophical concepts. The Saint-Simonians, influential in the building of the Suez Canal, believed that an elite of scientists, artisans, and other "productive" people should lead the state.[103] The fact that the resemblance of the Canal Zone to Bellamy's utopian blueprint was unintentional—no one had set up the infrastructure while flipping through the pages of *Looking Backward*—made the case of the Panama authors even more convincing: It added credibility to the argument that their vision was an almost natural evolution of the status quo. During the New Deal, when collectivist models resurfaced, a more active approach was taken to implement Bellamy's ideas. The engineer, educator, and utopian Arthur Morgan, biographer of Bellamy and commissioner of the Tennessee Valley Authority (TVA), applied concepts from *Looking Backward* to the workers' communities at the dam building sites.[104] Like the Panama authors, Morgan viewed the construction project as a testing ground to explore new planning ideas that could then be employed in the whole country.[105]

Benevolent Despot: The Rule of the Chief Engineer

No account of life in the Canal Zone would have been complete without a portrayal of its ruler, Colonel George W. Goethals (fig. 21). A major element of Roosevelt's swift overhaul of the Canal administration in 1907 was the appointment of Goethals as both chief engineer and chairman of the third Canal Commission. Although by law he was "no more than the first among equals"[106]—the commission consisted of seven men—Goethals acted and was perceived as if he had the powers of a dictator. In this role, he earned the admiration of his subjects. The storytellers called Goethals's regime "benevolent despotism, the sort of government that philosophers agree would be ideal if the benevolence of the despot could only be assured invariably and eternally."[107] The army officer, promoted to the rank of colonel in 1909, "might be classed as the most absolute despot on earth, although a benevolent one," another author wrote.[108] There was hardly a superlative with regard to Goethals that the storytellers would miss. Managing the Canal building and its workforce, "a huge, smoothly working engine of the highest capacity and efficiency,"[109] the chief engineer took the part of "the man at the lever,"[110] "the hero, the big brother, the father confessor, of every man on the job,"[111] "the genius of the Canal,"[112] "the blue-eyed czar at Culebra,"[113] "an Omnipotent, Omniscient, Omnipresent ruler."[114]

George Washington Goethals was born in 1858 in Brooklyn. He graduated from West Point second in his class and served his entire career in the U.S. Army, first as an engineer and instructor, later as a member of the General Staff in Washington, D.C. During his time in the Corps of Engineers, Goethals had gained some experience in building canals, locks, and dams.[115] According to the Panama Canal Act of 1912, which provided the framework for the administration of the Canal after its opening in 1914, Goethals became the first real governor on the Isthmus. The old position had been abolished after his appointment in

Figure 21. Chief Engineer George W. Goethals at his desk (reprinted with permission from Keystone-Mast Collection, UCR/California Museum of Photography, University of California, Riverside).

1907.[116] An alternative model granting the Canal Zone more autonomy was turned down. The permanent form of government on the Isthmus remained autocratic and was in many ways comparable to the status of the District of Columbia in the United States.[117] Goethals retained the position until 1916, when he was named quartermaster general in Washington to oversee the procurement for the army during World War I. Afterward he parted with the government and founded an engineering consultancy on Wall Street. He died in 1928.

The Panama authors were impressed by Goethals's administrative talent and his personality, characterized by both manly resolution and kindness, the classical traits of a benevolent despot: "His is a splendidly virile face, strong, kindly, and rigorously intellectual."[118] Female observers such as Mary McCarty were also impressed by Goethals, "the famous and popular autocrat with his blue eyes, white hair, tanned skin, courteous manners and agreeable laugh."[119] The colonel was famous for his discipline and knowledge of details, and he was always quick to react when problems surfaced.[120] Occasionally, he was accused of arrogance and megalomania—for instance, by Gorgas's wife, Marie.[121] Goethals did not have much respect for the work of the sanitation officer, who was his equal by military rank.

To most writers, Goethals personified the success story on the Isthmus. Although the groundwork had been laid before his appointment, the visible progress on the construction and infrastructure occurred during Goethals's reign. His leadership had proven superior to the red tape and inertia of the first Canal Commission and the lack of endurance of the second. This dictatorship became one of the reasons why the government had succeeded in Panama, and became part of the utopian vision that the writers construed for American society. In *Looking Backward,* Bellamy had also used a military figure to illustrate his authoritarian concept of the nation "under one control": Helmuth Graf von Moltke, chief of staff of the Prussian army in the war of 1870–1871 between Germany and France. "The effectiveness of the working force of a nation, under the myriad-headed leadership of private capital, even if the leaders were not mutual enemies, as compared with that which it attains under a single head, may be likened to the military efficiency of a mob, or a horde of barbarians with a thousand petty chiefs, as compared with that of a disciplined army under one general—such a fighting machine, for example, as the German army in the time of Von Moltke."[122]

Roosevelt admired the same qualities in his appointee: "Colonel Goethals has succeeded in instilling into the men under him a spirit which elsewhere has been found only in a few victorious armies," the former president wrote in his memoirs.[123] Because Goethals was an army man, his approach differed greatly from the management style of railroad managers and other businessmen. He had never held a position in industry. In his job and profession, he did not need to make a profit. Once again, Panama author Bullard cited a socialist who remarked that Goethals "won't make any more money if he gouges us. He don't increase his income by neglecting to put a guard on a machine. There isn't any money in it for him to have me living in a stinking tenement or eating bum grub. He can afford to be decent."[124]

Goethals wanted to present the Canal work as his personal achievement rather than the result of an experiment in new social practices. "Many have pointed to the Canal Zone as an example of a socialistic community," he told his audience at the Panama-Pacific International Exposition in San Francisco on "Goethals Day," September 7, 1915. "It was nothing of the sort—it is not now and it never has been. It was and is a perfect example of an autocratic government."[125] Goethals interpreted the cooperative society on the Isthmus not as an ideological choice but as "the result of a process of development."[126] For him, authoritarian rule had made the difference in Panama.

For his kingdom on the Isthmus, Goethals had developed a peculiar style of governance. He often oversaw the work from his personal railway car nicknamed "Yellow Peril."[127] Military rituals did not appeal to him. He never wore a uniform in public.[128] Instead, Goethals established customs that could very well have suited the courts of the traditional monarchs in history. Every Sunday, he received his "subjects" for an audience in his office in Culebra, where they could report their personal worries and complaints. "Everything, from the building and fortifying of the Canal, to explaining to Mrs. Jones why Mrs. Smith, whose husband gets twenty dollars less salary a month than hers, has received two more salt-cellars and an extra rocking-chair from the district quartermaster, rests on his shoulder, and he bears it all with a smile," one author reported.[129] The line "Tell the Colonel"—potentially an encouragement to denounce others—became the refrain for a song that was popular on the Isthmus.[130] Goethals's intent was obvious: to reinforce his executive with judicial power and let his presence be felt throughout the Canal Zone, even if he was not around. "The scarlet threads of his life-touch appear everywhere in the fabric, and nowhere more clearly than in this little, unconstituted Court of Caesar, which to the canal force at least has come to have far more importance than the whole judiciary system of the Zone, local, district, and supreme."[131] Historical comparisons were easy to find: the storytellers likened Goethals to the charismatic caliph Harun ar-Rashid[132] and the Venetian doges[133] and called him "The Solomon of the Isthmus." The chief engineer seemed pleased with his image of the wise potentate and happily admitted that "I became the father confessor."[134]

The iconography of Goethals's depiction was not exceptional. Prior to his appointment as secretary of war in 1904, Taft had attended a costume ball in the Philippines dressed as a Venetian doge.[135] Taft served as the first civilian governor of the new American colony and head of the Philippine Commission, a counterweight to the military administration of the island group, which had become part of the U.S. empire in 1899 after the Spanish-American War.[136] After more than two years of brutal warfare, the American military consolidated their hold on the country by striking down a guerilla independence movement and capturing their leader. Simultaneously, starting in 1900, the government in Washington, D.C., pursued a policy of "benevolent assimilation," tainted by racism and designed "primarily to remake the colony in the image of the United States,"[137] which Taft, a former judge, and other members of the civilian commission were to carry out. The objectives were to prepare the country with 7.5 million inhabitants for self-governance, install a comprehensive infrastructure for primary education, and develop the economy. According to historian Glenn Anthony May, execution was poor in all three areas.[138]

It is important to note the differences between the colonial venture in the Philippines and the building of the Panama Canal. Both projects shared the racist and gendered ridicule of the local population, and both were essentially technological projects: In the Philippines, the task was to be solved through social engineering.[139] While the Filipinos were to receive the alleged blessings of American civilization, the process in Panama went the other way. Excluding the Natives and West Indian workers from their interpretations, the Panama authors viewed the Canal Zone essentially as an American state and a model *for* the United States—not vice versa. The building of the Panama Canal fascinated policymakers and the public, whereas the Philippines failed to attract the attention of even the members of Congress: Few of them traveled there, and their decision making regarding the colony was solely motivated by short-term political dynamics in Washington.[140]

The failure of the Philippine venture may have changed the course of the American empire proper—the formal empire accumulated by annexation—if empire is viewed solely in terms of foreign relations. For the expansionists, however, the foreign and the domestic were inseparable, and the building of the Panama Canal is perhaps the best illustration of this view. The doge motif carried different meanings for Goethals and Taft: while the Philippine governor played the part of the benign patron for the Natives, Goethals embodied a new kind of American leader who combined the traditional and modern qualities of monarch, soldier, and engineer.

The fact that Goethals was an engineer contributed significantly to his positive image. In all industrial nations, engineers were increasingly identified as modern leaders.[141] Journalists and writers "staged the engineer as a male cultural hero,"[142] who became the protagonist of epics such as Kellermann's fictional account of the building of a transatlantic tunnel. In the decades prior to and during World War I, when it became apparent that technological change affected all of society, civil engineers seemed ideally suited to confront this challenge. "It is to the fine scientific habit of mind, with its catholicity of interest, its reverence for facts, its high sense of the value of human life, that the country must look for its salvation," the journalist Ray Stannard Baker asserted in a portrait of Frederick W. Taylor, the efficiency expert who was about to become the most prominent representative of the new engineer-manager type.[143] As industrialization progressed, the education and employment of engineers began to reflect the changes in the business world. After the Civil War, the traditional method of on-the-job training in machine shops was slowly replaced by a more theoretical and integrated education at engineering schools. Many graduates were recruited by corporations that had big administrative and management departments and were constantly aiming to increase productivity. Between 1900 and 1930, the number of engineers in the United States increased dramatically from 45,000 to 230,000 and yet accounted only for 0.5 percent of the total workforce. According to historian David Noble, over 90 percent were employed as technical workers and managers in industry, three quarters had a middle-class background, and all but one in a thousand engineers were male.[144]

In his *Principles of Scientific Management* (1911), Taylor presented rigorous methods of making mass production more efficient, especially in areas where machines and human labor interacted. Employing stopwatches to measure workflows and prevent "soldiering," Taylor frequently met resistance from labor representatives. His ideas did have a varying impact on corporations and their assembly lines but resonated all across the world—even in the Soviet Union.[145] Taylor also prompted a debate, a movement even, among middle-class reformers. While companies were mainly interested in gaining a competitive advantage—often at the expense of the workers—Taylor's intellectual followers intended to make society as a whole more efficient and ultimately more humane. The villains in this scheme, as in Roosevelt's and Bellamy's, were the materialistic and egocentric capitalists. The movement took different shapes, such as the New Machine association, which promised to "take control of the huge and delicate apparatus of industry out of the hands of idlers and wastrels."[146]

The engineer was the perfect leader to transcend the interests of financiers and laborers and serve the public good. There is probably no single project and no single figure in American history that illustrates this belief better than the Panama Canal and its chief engineer, Goethals. As an employee of the government, not a private corporation, he came to represent the "pure" image of the engineer, detached from business interests. He was a military officer, not a profiteer; resolute, yet equipped with civilized manners. In a poem, he was even called the "prophet-engineer," bringing redemption from the public sins of the past.[147] In a speech before the international Engineering Congress at the Panama-Pacific International Exposition, Goethals interpreted his own role along the same lines and provided the rationale for the benevolent dictatorship in the Canal Zone: "The Canal is another illustration of the functions of the engineer, his uses of the forces and materials of nature for the benefit of man. It is another instance of the fact that engineers are fitted for great executive and administrative functions, and also that they can accomplish and manage a government for the satisfaction of those governed. Believing in straightforward, practical administration, as the engineer does, such a government to be successful must not be political but autocratic, for with politics involved there would enter an unfamiliar factor opposed to the engineer's training and ideals, and he would fail."[148]

Goethals's authoritarian management style is illustrated by the authors' description of labor conflicts on the Isthmus. Although the Canal Zone was an open shop, many skilled workers were organized in trade unions. A strike, however, was not a weapon they could use to put pressure on the Canal Commission. When boilermakers in Gorgona and Empire demanded a wage increase in November 1910, Goethals ordered their deportation to the United States without any further negotiations.[149] This was legal since the Canal Zone was not a de jure part of the United States.[150] The "Solomon of the Isthmus" even refused to negotiate with committees representing groups of laborers. After such an appeal, he noted: "I also informed the committee that it was ill-advised to make *demands* for increases in pay or other concessions, and thereafter none such would be given any consideration. Requests, if properly made, would be received and acted on according to their merits."[151] The only demand the workers could make was to appeal to the wisdom of the despot. Goethals faced another challenge when laborers announced they would go on strike in February 1911 after a locomotive engineer had been severely (and in their mind too harshly) punished following a work accident. This time Goethals's mere threat to dismiss and deport the laborers sufficed: they abandoned their protest.[152] In general, the Panama authors refused to pay much attention to controversial labor issues. In their opinion, there were no reasons for strikes since the working conditions—like everything else in the Canal Zone—stood on a "utopian basis."[153]

Goethals's right hand, Commission secretary Joseph Bucklin Bishop, helped create the positive image of his superior. In his own book on the Canal construction, he cited a distinguished (yet anonymous) engineer who found great praise for the project: "I have never seen its superior—such perfect co-ordination and such energetic prosecution at every point, all under absolute control."[154] But the dictatorship also prompted critics to raise their voices. One author, who called Bishop's newspaper the *Canal Record* "the weekly bulletin of the despotism," admitted that no one seemed to protest the abolishment of democratic decision making, yet he remained doubtful. "The laws are made in Washington and Culebra, without question as to the wishes of the people, and there is a consequent loss of social development. . . . I say this Government has been too kind, because no matter how pleasant it is to have others do one's thinking the effect of five years or more of benevolent despotism in the Canal Zone, has convinced me thoroughly of the educative value of a democratic form of government."[155]

The despotic regime on the Isthmus, symbolized by the persona of Colonel Goethals, became an important element in the utopian interpretations of the Panama Canal. Whereas the Erie Canal had been celebrated as a collective yet democratic project, "undirected by absolute authority,"[156] the new era seemed to require not consensus and compromise through equal participation but only the benevolence of the state and its experts. Equipped to cure the ills of industrialization and over-civilization, the engineer-soldier became the role model for the future statesman: efficient, patriotic, just, and successful—a trademark signifying the concentration of social forces in the hands of the government.

Drift and Mastery: The Canal, the New Intellectuals, and the War

Interpreting the Canal Zone as a model for the United States, the Panama authors built on social visions that had permeated American society for many decades. The storytellers who had been involved in the engineering project and the journalists who had come to the Isthmus to write a potentially bestselling book may or may not have been conscious of the influences directing their observations. Another group of writers, however, deliberately chose to integrate the lesson of Panama into a politically charged program for the future America. Most of these commentators, all of whom had some kind of attachment to the "Progressive movement," did not muster the fascination for details necessary to complete a book-length study on the Canal building, but they read the works of their fellow Panama authors and then drew their own conclusions in magazine articles, or they included references to the Canal in general analyses of the American condition. A closer look at this group of authors will reveal the complex relationship between the Canal discourse and the evolving debate on the role of the state prior to and during World War I.

Ray Stannard Baker (1870–1946) was one of these writers. His article "The Glory of Panama," which appeared in the *American Magazine* in late 1913, provides perhaps the best summary of what the storytellers had seen on the Isthmus.[157] "I agree beforehand not to say a word about the amount of red earth taken from Culebra Cut, nor try to convey the unconveyable by showing how many times around the earth the barrels of cement used in the Gatun locks would reach if placed end to end," Baker noted with a twist of irony (33). The engineering feat itself did not interest him—it was "The Spiritual Factor" (34) of the work and of the Canal society that held the greatest implications for Baker: "If the canal should be destroyed the day after it is finished it would still be worth all it cost us" (33).

Baker had been an editor of the *American Magazine* for many years, even though he was better known for his affiliation with *McClure's Magazine*, one of the cheap, mass-circulation publications that redefined journalism at the turn of the twentieth century. Through investigative reporting and easy-to-read, illustrated articles, written in an accusing tenor and with great attention to detail, authors such as Baker, Ida N. Tarbell, and Lincoln Steffens had examined the dark sides of industrialization and brought to light the abominable living conditions of the urban poor and the manifold discriminations against racial minorities. President Roosevelt, though generally sympathetic with the enemies of a rampant capitalism, had named this group of authors the "muckrakers," a term that expressed his ambivalence whether the fearless writers had not gone too far with their claims.[158] As discussed earlier, the Canal project had been the target of muckraker Poultney Bigelow, whose conclusions were heavily criticized by members of the government (and later by the Panama authors) and prompted a Senate investigation into the situation on the Isthmus before the Goethals regime was established.

By 1913, when the excitement over the journalists' exposures and their new reporting style had subsided, typical muckraking articles appeared less frequently. Baker, who had earned his reputation through the account of a mineworkers' strike in Colorado and an influential essay on African American citizenship, was now regarded as an established author. Following World War I, he was named the head of President Wilson's press office and later became the politician's official biographer.

To convince his readers of the significance of the Canal project, Baker used his credibility as a former muckraker and described the living and working conditions he had encountered on the Isthmus. He noted that in contrast to most of industry, which had shown "little care whether the workers lived like pigs or died like flies," the organization on the Isthmus had achieved its success by ensuring the social welfare of its laborers, "based primarily upon recognition of the fact that the task is a *public*, not a *private*, enterprise, that the work is being done not by mere employees, but by citizens and free men. Slaves may be driven or forced, citizens must be inspired" (33). Baker recounted Gorgas's sanitation efforts and the victory over yellow fever and malaria, and he described the benefits the Canal workers enjoyed with regard to housing, meals, and entertainment. He noted that the laborers probably earned the highest wages in the world due to the fact that "percentages which in private enterprises go into interests and profits may here be distributed to the workers" (34). Unlike other, usually racist Panama authors who happily acknowledged the differences between the gold and silver workers, Baker stressed again and again, almost imploringly, that "[n]o distinctions are made between white, yellow, and black men" (34). Whether he actually believed this can only be guessed.

Baker paid tribute to the use of technology in the Canal Zone, exemplified by bread-, ice- and coffee-making plants as well as labor-saving inventions for the control of lock gates or for laying railroad tracks. He welcomed the transparency of the project achieved through the publicity of commission secretary and *Canal Record* editor Bishop and the absence of favoritism. Baker assigned the responsibility for this particular style of government to Chief Engineer Goethals, whom he had described in an earlier essay in the *American Magazine.* "No pull goes with the old man," he quoted a young foreman as saying (35). Goethals's method of constantly improving the workflow had eliminated waste: "Efficiency is the watchword—maximum results with minimum expenses," Baker concluded, invoking Taylor's *Principles of Scientific Management.* As a consequence, the Canal would be finished a year ahead of time for $8 to $10 million less than the $375 million estimate "made soon after Goethals went to Panama" (36). Baker presented the chief engineer as the new industrial statesman, replacing the tycoons of the early industrial age: "Rockefeller and Morgan are vanishing types of leadership: and their methods have already passed into disfavor" (36).

Finally, Baker turned to the success story of the Canal, spanning ten years from a difficult start, during which officials were "doubtful of our own abilities, not trusting the new attitude toward public work" (33), to the realization he witnessed, and he created a parable of national metamorphosis: "In a literal sense the Canal is changing the viewpoint of the nation itself, an effect which will be still more noticeable with the passage of time. It may well be that historians of the future will mark the date of the canal as the beginning of a new epoch in American life. As the Spanish War gave us outside interests and a new confidence in ourselves as a factor in the politics of the world, so the successful completion of the canal is giving us a new confidence in the ability as a nation, rightly led, to do things of itself which in the past have been left to private enterprise. We have gained self-confidence. We shall hesitate less in the future about national undertakings, stupendous and difficult though they may appear" (36).

There were other public intellectuals who—like Baker—wrote about general political issues and at the same time displayed an interest in the Canal. Arthur Bullard (1879–1929) was a case in point. The author of *Panama: The Canal, the Country and the People,* which had been published in an earlier version under his pen name Albert Edwards, was a journalist and also the author of *Comrade Yetta* (1913), a novel on reform movements.[159] Before and after the Soviet revolution, Bullard reported for American magazines from Russia. Despite his socialist leanings, he became involved in Wilson's propaganda agency, the Committee on Public Information (CPI) founded during World War I, and was later appointed head of the State Department's Russian Division. Many radical intellectuals went through a similar process of assimilation.

The most influential group of this time period is sometimes referred to as the Young Americans or New Intellectuals (or even Young Intellectuals) and included Herbert Croly, Walter Lippmann, Walter Weyl, William English Walling—a friend of Bullard's—Randolph Bourne, Van Wyck Brooks, and Waldo Frank. The fact that the public weight of these authors was significantly diminished during the 1920s may account for the lesser role they have played in the historiography of American thought, although scholarly interest has never completely subsided.[160] Prior to World War I, the debate among them on the expanded role of the state—sometimes summarized as "state socialism"—was at its height. Which role did the Panama Canal project play in this argument? To what degree did the opinions of the intellectuals exert an influence in the political realm? How did the war affect the debate and its participants? A review of the Young Americans will also shed light on the question why the Panama discourse lost much of its relevance shortly after the completion of the Canal.

It is tempting to assign the term "Progressive" to these authors, yet the question remains: what can be gained from such nomenclature? As discussed before, the "Progressive movement" referred to a diverse group of actors that, depending on the definition, included figures as incommensurable as Theodore Roosevelt and Woodrow Wilson, social workers, muckrakers, businessmen, conservationists, and feminists. The term was most widely used during Roosevelt's unsuccessful attempt to regain the presidency in the Bull Moose campaign of 1912 as the spearhead of the Progressive Party. But rather than identifying very specific ideas and people with the label "Progressive," it makes perhaps more sense to describe the "movement" as an ongoing discourse, shaped largely by middle-class actors in their search for order, as the attempt to construct a new America in light of the changes that industrialization and technology had generated on all levels of society. "We are 'emancipated' from an ordered world. We drift," the journalist Lippmann wrote.[161] The task was to regain control. There may be no better illustration of this spirit than Baker's "The Glory of Panama." Even though the state on the Isthmus was unique, its interpretation was linked to the cultural developments at home. "It may be truly said that Goethals is succeeding at Panama because the country to-day is coming to be full of little Goethalses," Baker wrote.[162] Due to this new sense of common effort, as opposed to the ragged individualism of a less complex world, the future role of the state became one of the central questions in the debate.

Most of the New Intellectuals were young and well educated and came from middle-class backgrounds. In their academic training, they had studied the works of the pragmatic philosophers William James and John Dewey. They admired European (often German) models of governmental power and discipline. Sometimes they were or had been self-described Socialists, but in a more mainstream sense than orthodox Marxists.[163] As "a kind of unconscious archetype,"[164] Bellamy and his writings represented another inspiration for these authors. The best label for the political style of this group, even though it has different connotations today, is "radical." The intellectuals discussed their views in books and magazines that—contrary to the writings of the Panama authors—were not aimed toward a mass market but were read and highly regarded by the political elite.

The debate on "state socialism" had its roots in the excesses of capitalism criticized by Roosevelt and socialist thinkers alike. Without a moral architecture, they argued, the economic system was unaware of the suffering it caused. "The industrial goal of the democracy is the socialization of industry," Walter Weyl (1873–1919), a trained economist, wrote. "The democracy seeks to attain these ends through government ownership of industry; through government regulation; through tax reform; through a moralization and reorganization in the interest of the industrially weak."[165] Recalling Baker's words, Weyl's colleague Walter Lippmann argued along the same lines: "The remedy of commercialism is collective organization in which the profiteer has given way to the industrial statesman. The incentive is not alone love of competent work but a desire to get greater social values out of human life."[166] The shape of this "collective organization" was yet to be determined. Whereas Bellamy had advocated the state as "the sole corporation" and "The Great Trust,"[167] the socialist, General Electric manager, and inventor Charles P. Steinmetz, a German immigrant, argued that a company such as the one he worked for could also assume this role.[168]

The range of the discussion demonstrates how strongly it was rooted in the realm of theory and wishful thinking, as opposed to political application. The Panama Canal, "undoubtedly the largest single undertaking by any government in any period,"[169] was therefore a more than welcome illustration of what the future America might look like. "Government goes into business," Weyl observed. "In the construction of the Panama Canal, the government builds roads and railroads and conducts dozens of separate enterprises."[170] Although impressed by the regime on the Isthmus, the Young Americans were also wary of its implications. "The construction of the Panama Canal is a classic example of what government can do if it is ready to centralize power and let it work without democratic interruption," Lippmann noted.[171] He seemed awed by the possibilities of an autocratic system and at the same time repelled by the consequences: "The bureaucratic dreams of the reformers often bear a striking resemblance to the honest fantasies of the utopians. What we are coming to call 'State Socialism' is in fact an attempt to impose a benevolent governing class on humanity."[172] In his earlier works, the writer William English Walling (1877–1936) outright rejected the arguments for a more powerful state, afraid that such a development would only strengthen the conservative, privileged groups within society: "It would seem that in the mind of most 'State Socialists' and social reformers the cure of 'poverty' and of the industrial inefficiency of the workers is united indissolubly with coercion, if not with military government."[173] Later, as the United States became involved in the war in Europe, Walling took a more positive view on state socialism.[174]

The intellectuals' shifting course between embracing and rejecting collectivism points to an inherent conflict. It reflected the evolution of their thinking, their constant struggle with textbook socialism and the idealism of the utopians on the one hand, and the limitations of democratic involvement in a modern society and the necessities of Realpolitik on the other. Their burdensome mission was to define liberalism in the new century. As the involvement of the intellectuals in government politics increased and the country became more and more entangled in World War I, their internal struggle intensified. Probably the most prominent representative of this intellectual meandering is Walter Lippmann (1889–1973).

A native New Yorker, Lippmann came from a well-to-do German-Jewish family. Educated at Harvard, he soon earned a reputation as a brilliant essayist.[175] With the publication of *Drift and Mastery* at the age of twenty-five, he parted with socialism and embraced the pragmatic political philosophy that was to become the trademark of the ambitious new magazine the *New Republic*. In 1914 Lippmann and Weyl joined forces with Herbert Croly (1869–1930), who due to his age could not really be considered a member of the Young Americans. Five years earlier, Croly had published *The Promise of American Life*, a book that was seen by many as the manifesto of the Progressive movement and a blueprint for Roosevelt's campaign for presidency. Croly advocated a strong centralized government, the eventual nationalization of large corporations, and a cautious expansion abroad. The book's direct influence on the public was negligible, as it sold only about 7,500 copies.[176] With the financial backing of the J. P. Morgan banker Willard Straight and his wealthy wife, Croly founded the *New Republic* (subtitled *A Journal of Opinion*), with Lippmann and Weyl serving as editors. It was a fortunate time for magazines focusing on political thought and cultural critique. At about the same time, the less solemn *Seven Arts* was started by Frank, Brooks, and Bourne. It circulated for only a year, when its backer, a wealthy woman who had been advised by her psychoanalyst to sponsor a radical magazine, succumbed to pressures of her family and discontinued her support.[177] The *New Review* (subtitle: *A Critical Survey of International Socialism*), first printed in 1913, had Walling on its board of editors; Bullard and Steinmetz served as advisers.

With the outbreak of the war in Europe in the summer of 1914, the editors of the *New Republic* had to take a position from the very beginning of the magazine, whose first issue appeared on November 7, 1914.[178] In this first issue, the intellectuals presented the conflict as a futile effort; men would "spend years learning to make war; they do not learn to govern themselves." At the same time, the respect for war as a collective task, "the one activity that men really plan for passionately on a national scale," could be read between the lines. Was the true unity not to be found in a "moral equivalent of war," as William James had suggested, achieved by the esprit de corps that accompanied it? "It requires a trained intelligence to realize that the building of the Panama Canal by the American Army is perhaps the greatest victory an army ever won," the editors suggested.[179] Nonetheless, the possibility of an authoritarian system, praised by the Panama authors with regard to the Canal project, still frightened them. "The collectivism we are seeking cannot be imposed with an iron fist and run by martial law. . . . It must be made by experiment, by argument, demonstration; it must be the work of a people that is training itself in cooperation."[180]

In the course of the following years, as the intellectuals discovered President Wilson as their new counterpart in politics, their opposition to the war began to crumble. Before the fruitless Bull Moose campaign of

1912, Roosevelt had been the sole hero of the Young Americans. "He was the first political leader of the American people to identify the national principle with an ideal of reform," Croly had written in *The Promise of American Life*.[181] But as the former president turned to game hunting and the Progressive Party seemed headed on a path toward irrelevance, the intellectuals' loyalty began to dissolve. Wilson, TR's Democratic opponent, demonstrated his determination as a reformer when he initiated the Federal Trade Commission Act and the Clayton Anti-Trust Act, both passed in 1914 and designed to put questionable business practices under scrutiny. Prior to his reelection in 1916, the president made an effort to court the editors of the *New Republic* and eventually gained their support. By the end of the year, they endorsed a limited U.S. participation in the war and the prospective establishment of a League of Nations.[182] There were two reasons for this decision: First, the sudden proximity to power flattered the writers. To Lippmann, a closer involvement with government matters may have seemed the logical next step in his career. Second, what if the war was, after all, the only chance to create an expanded state for the public benefit? Not only the editors at the *New Republic* were asking themselves this question. "Internationally the war is bound to modify national individualism in favor of federation of nations; nationally the war strikes a powerful blow, perhaps the final blow, at the decrepit system of economic individualism," a Marxist author for the *New Review* had suggested shortly after the passenger liner *Lusitania* had been sunk by a German submarine off the coast of Ireland and the U.S. entry into the war seemed closer than ever before.[183]

Wilson had won his reelection with the help of the slogan "He Kept Us Out of War," but after the declaration of unlimited submarine warfare by Germany, American involvement seemed inevitable. Finally, on April 2, 1917, the president asked Congress for a declaration of war "to make the world safe for democracy." This meant that mobilization had to begin quickly and forcefully, aided by a war economy that would raise the experiment of collectivism carried out in the construction of the Panama Canal to a new level. "Contemporaries of all persuasions regarded mobilization as a testing ground for the principle of government involvement in the economic life of the nation," historian David Kennedy notes.[184]

Faced with conscription, Lippmann wrote a letter to Secretary of War Newton D. Baker asking for an exemption. Foregoing his journalistic independence, he joined Baker's staff and in the fall of 1917 was appointed general-secretary of a secret think tank named "The Inquiry," whose task was to draft the Fourteen Points, Wilson's famously lofty program for peace.[185] Like Bullard and others, Lippmann now worked *for* the government—to fulfill his patriotic duties and play an active part in the advancement of his own ideas. Not all writers approved of these entanglements. Observing his fellow intellectuals, Randolph Bourne (1886–1918), who had remained an outsider among the Young Americans, noted "how soon their 'mastery' becomes 'drift,' tangled in the fatal drive toward victory as its own end, how soon they become mere agents and expositors of forces as they are."[186] Judged in hindsight, his words seemed almost prophetic.

Wilson carried out a number of measures that at first seemed to confirm the intellectuals' notion that the war would catapult the country into a collectivist era. But upon closer examination, they turned out to be short-lived compromises designed to drill the American economy into shape for the war. By design, the new War Industries Board (WIB) established in August 1917 had the task of controlling prices and redirecting production in order to ensure war supplies. In reality, the price-fixings were gentlemen's agreements between the government and big business—made to work, not to change, the system. Businessmen became involved in almost all of the new agencies, as their managerial experience was badly needed to make the slow-moving government bodies beat to the fast pulse of war.[187] Bernard Baruch, a Wall Street speculator, was appointed chairman of the WIB. Not soldiers but figures such as Baruch and Food Administrator Herbert Hoover, a later president, became the most popular figures of the American war effort.[188] Shortly after the armistice on November 11, 1918, the WIB was dismantled. The railroads, perhaps the strongest symbol of government involvement in the economy (or its absence), had been placed under control of the Railroad Administration by the end of 1917 to bring the surge in fare prices to a stop. However, business as usual returned with the Transportation Act of 1920.[189] A "U.S. Railroad" remained limited to the Canal Zone. While the government and the business sector profited from the war measures in terms of better contacts, efficiency, and, most important of all, standardization—processes that had already been under way—little of the anti-business rhetoric of the Young Americans resounded in Wilson's policies.[190] The president had chosen the "safe middle way."[191]

Harsher consequences of the participation in the war could be felt in other areas. The Committee on Public Information, headed by journalist George Creel, evolved into a full-fledged propaganda machine. With the help of the Espionage Act of 1917 and the Sedition Act of 1918, the Justice Department began not only to prosecute friends of the German Kaiser but also members of left-wing groups who opposed the war, such as the radical labor union Industrial Workers of the World (IWW), whose members were nicknamed "Wobblies." Socialist leader Debs received a ten-year sentence and was put in a federal prison, which did not prevent him from winning close to 920,000 votes in the 1920 presidential election—slightly more than he had received in 1912.[192] Lippmann made his own acquaintance with the spirit of the times in Europe, where he had to compose leaflets to be dropped behind enemy lines.[193] Censorship and other excesses of propaganda proved deeply troubling to the Young Americans, who not long ago had advocated an expanded state to be created "by experiment, by argument" instead of by the "iron fist."

The manifold disappointments of the New Intellectuals—many of them former admirers of German culture—were aggravated by the Treaty of Versailles, which the *New Republic* called "a punic peace of annihilation."[194] The Senate refused to endorse Wilson's main contribution to the treaty, the League of Nations. But it was not only the failure of the politicians (and their own failure) to lift the country into a new era of public spirit that bothered the authors/advisers. They had lost faith in the possibility of a national community. Modern democracy, it dawned on Lippmann, could not live up to its promise. In a complex world, communication between the political leaders and the broader public was always bound to involve opportunism and stereotypes. The book *Public Opinion*, published in 1922, was another step in Lippmann's intellectual evolution toward skepticism.[195] To assume that "somehow mysteriously there exists in the hearts of men a knowledge of the world," of "the perplexities of government and industry," was an illusion, he argued.[196] Political decisions could not be expected to be rational. "The world that we have to deal with politically is out of reach, out of sight, out of mind. It has to be explored, reported, and imagined."[197] He was no longer willing to entrust his own profession, journalism, with such a difficult if not impossible task. Half-heartedly, Lippmann suggested that an organization of experts assist politicians and the media in making "the pictures inside people's heads" conform to reality.[198] The shattered belief that resolution combined with reform could achieve a new civic unity—that the state and the individual could actually become *one*—marked the end of the Progressive age.

World War I also diminished the enthusiasm the Panama authors had expressed for the Canal. Disturbances caused by the German submarine warfare and the decline in international trade meant that immediately after its opening (and for many years to come) the engineering feat did not meet the economic expectations. Moreover, massive slides in Culebra Cut in the fall of 1915 resulted in a closure of the Canal for big ships that lasted for several months.[199] For the most part of that year, however, the interpretations of the storytellers did resurface within the Unites States at the Panama-Pacific International Exposition in San Francisco—a world's fair held to celebrate the achievement of the waterway. In many ways, the exposition (discussed in chapter 5) was the culmination of the Canal discourse. Afterward, the vision constructed on the Isthmus of Panama lost its immediate hold on the American middle class. The Canal's meanings had almost literally been washed away after its completion. Twisting Baker's argument that the Canal would still be worth all it had cost even if it were destroyed the day after it was finished, it could be argued that the completed Canal was worth no more than the tolls paid for its passage but had been destroyed as a working symbol of the American future.

While the perception of modernizing processes and their effects on the state and the individual changed in the 1920s, the processes continued. During the Depression in the 1930s and through the policies of the New Deal—which Lippmann, incidentally, opposed—the discourse on collectivism resumed its dominating role in American society. At the same time, the authoritarianism of the emerging fascist regimes in Europe indicated that the utopian visions developed in Panama had been but foreshadows of things to come. Abbot, one of the most popular Panama authors, reflected in his memoirs on an encounter with the Italian dictator Benito Mussolini. Writing in 1933, Abbot was intrigued by fascist Italy, which may have reminded him of the benevolent despotism he had described in Panama. He praised a system that promised freedom "from the tyranny of the lawless, for exemption from the demands, often unreasonable and extortionate, of trade unions, for liberation from the bonds of official and traditional red tape, and for emancipation from the bondage of moldy tradition and worn-out theories."[200]

Abbot's comments demonstrate how close the Canal vision could have come to totalitarianism.

Summing Up: The "Happy Dream" of the American Future

In the jungle of the Isthmus, the Panama authors had stumbled across utopia. Most of the writers who traveled there did not encounter a foreign country, or a segregated society, but an American state that seemed distant and strangely familiar at the same time—like a glimpse into the future. With blinding racism and from a tourist perspective, they shut the West Indian laborers out of their view of the Canal community and were left with the healthy, white, and mostly male inhabitants of El Dorado. The government had succeeded private companies as an employer and provider of social services. The profit motive had been replaced by efficiency and idealism. Technology ensured that everyone was provided for. In the suburban landscape of the Canal Zone, the hustle-and-bustle environment of the industrialized cities had succumbed to a morally cleansed lifestyle of consumption. "After years of hearing of the shame of corrupt politics and of inhumanity of industry in America," Baker wrote, "it is refreshing, indeed, to find here not only an exemplification of the ancient fibre of the race but a realization of its newest ideals."[201]

The interpretations of the Panama authors were directed at a middle-class audience eager to make sense of the new century and to overcome the divisions within society. Therefore, they decided not to present the Canal Zone as a socialistic experiment reeking of working-class rule. There were other patterns in the Canal project that their readers would recognize: Edward Bellamy's utopia, whether it was explicitly mentioned or hovering in the back of the authors' minds, was their blueprint for the construction of a new America on the Isthmus. Like Bellamy, the authors were willing to let go of democratic traditions. In an ideal state, a benevolent despot such as Goethals or an elite of experts was sufficient to preserve the status quo. The engineer-soldier became the model of the new social and political leader.

There had always been arguments, sometimes raised by the authors themselves, that it would be folly to apply the lessons of Panama to the United States. "If Government abattoirs could sell meat at lower rates

to the people of any State than to the people of every other State in
the Union, we should have a real parallel to Panama. And if popular
elections would always give us a Goethals for the post of construction
engineer and a Gorgas for sanitary engineer, the parallel would be still
closer," a writer for the *Nation* concluded mockingly in a response to
Baker's "Glory of Panama."[202] And yet, in most of the writings, the parable of the Canal Zone proved stronger than the plain facts. It was the
better story, the national epic: on the Isthmus, America had completed
its evolution and reached the end of a process, the "achieved condition
of order and refinement," as Raymond Williams phrased his definition
of civilization.[203] All that was left to do was to try it at home.

This utopia was a paralyzed paradise, filled with the fear of contin-
gency and rapid change. In his essay *Drift and Mastery*, Walter Lippmann
pointed out the shortcomings of what he called the "honest fantasies of
the utopians" (among them, Bellamy):

> Life is fixed: the notion of change is rare, for men do not easily associate
> perfection with movement. Moreover, the citizens of these utopias are
> the disciplined servants of the community. They are rigorously planned
> types with sharply defined careers laid out for them from birth to death.
> A real man would regard this ideal life as an unmitigated tyranny. But
> why are the utopias tyrannical? I imagine it is because the dreamer's
> notion of perfection is a place where everything and everybody is the
> puppet of his will. In a happy dream the dreamer is omnipotent: that is
> why it is a happy dream. So utopias tend toward a scrupulous order,
> eating in common mess halls, mating by order of the state, working as
> the servant of the community. There is no democracy in a utopia, no
> willingness to allow intractable human beings the pleasure of going to
> the devil in their own way.[204]

The New Intellectuals probably realized that the vision expressed in the
books and articles of the Panama authors was severely flawed. At the
same time, they were driven by similar longings for a new public spirit
and admired the war-like accomplishment of the Panama Canal. The
project on the Isthmus became the key witness for their political agenda.
Overriding their reservations toward authoritarian collectivism, they
put their hopes on the once-in-a-lifetime chance that the real war seemed
to offer for the introduction of a centralized state. When it did not materialize, their faith in the benign nature of American democracy was
shattered. Repelled by the inherent tyranny of the utopians' visions as
well as Wilson's wartime propaganda, and unwilling to believe in the

public's emancipation from populist influences, the intellectuals continued to struggle with the question of governance and, ultimately, democracy in modern society.

In the minds of the middle-class citizens who believed the lessons of Panama, the United States had come close to fulfilling its destiny on the Isthmus. In contrast to other countries undergoing the same transformations, the American nation was seen by many as an experiment designed to evolve into Lippmann's "happy dream." In 1892, four hundred years after Columbus's first voyage to the New World, Edward Bellamy had written to his supporters: "We are today confronted by portentous indications in the conditions of American industry, society and politics that this great experiment, on which the last hope of the race depends, is to prove, like all former experiments, a disastrous failure. Let us bear in mind that, if it be a failure, it will be a final failure. There are no more new worlds to be discovered, no fresh continents to offer virgin fields for new ventures."[205] In this sense, the Canal Zone was not a new country or society but a laboratory for the unfinished American experiment.

5

Celebrating the Canal

The Panama-Pacific International Exposition

The success stories of the Panama authors had all been written with a happy ending in mind: the grand opening of the waterway. But political developments in Europe interfered, and the event turned into an anti-climax. On June 28, 1914, a Serb nationalist killed the heir to the throne of Austria-Hungary and set off a chain reaction. By August, the European powers had mobilized their troops and Germany invaded Belgium. Public attention in the United States shifted to the evolving World War I. The great Canal parade of battleships, freighters, and luxury yachts under the command of President Wilson, which storyteller Farnham Bishop had imagined two years before, never took place.[1] The first official ship passing through the Panama Canal on August 15, only eleven months after the digging of Culebra Cut had been completed, was the freight boat *Ancon,* a simple transport vessel for cement.

A few months later, the opening of the Panama Canal once again made headlines, albeit in a different arena. The authors interpreting the events on the Isthmus had unfolded a vision not of a foreign territory but of the future American society. In this sense, it was appropriate that the Canal opening was embodied not by the actual passage of ships but in the abstract realm of a world's fair within the United States, the Panama-Pacific International Exposition (PPIE) in San Francisco. The admission figures of close to nineteen million visitors sound impressive, but most likely they included many multiple entries.[2] As before in the case of the Canal, most Americans learned about the fair in hundreds of articles and guidebooks and by looking at photographs and postcards of its attractions. The meanings of the exposition were translated to a middle-class audience through the lens of interpretation.

It is a remarkable coincidence in the historiography of the Panama Canal that the cultural projects on the Isthmus and in San Francisco have rarely been discussed together.[3] Instead, they have been treated as events in separate historical disciplines, even though contemporary authors often concluded their Canal books with an outlook on the exposition.[4] This chapter is based on the assumption that the PPIE must be understood in the context of its actual purpose, the celebration of the Canal endeavor. It was also embedded in two other discourses, which have been the foci of scholarly discussion: the history of world's fairs and the regional history of San Francisco. The exposition was the first (and, arguably, the final) test of the collective visions that policymakers and authors had constructed on the Isthmus. It took place in a real American city, but, like the interpretation of the Canal, it was confined to the sphere of dreams, display, and imagination.

Progress and Amusement:
World's Fairs and Their Messages

The world's fairs held in the United States and Europe are a phenomenon of the industrial age. The first exposition, the Crystal Palace Exhibition, took place in London in 1851. The Eiffel Tower in Paris, built for the exposition in 1889, remains the most famous symbol of these events, which generally left few architectural traces in the cities where they where held. They were ephemeral, visionary incorporations of progress and expansion. Four big expositions took place in the United States, the fairs in Philadelphia in 1876, Chicago in 1893, St. Louis in 1904, and San Francisco in 1915. The most influential of these was the Columbian Exposition in Chicago, setting the stage for the grand exposition themes in the age of empire and offering a preconceived "nationalizing synthesis"[5] of white civilization and territorial expansion. The Chicago fair was also a prototype for subsequent fairs in terms of exposition layout, architecture, and exhibits. This means that despite all the progress that was made in the meantime, much of what was shown in San Francisco (and especially how it was shown) was very similar to the fairs in Chicago and St. Louis.

Although world's fairs were important in international, national, and regional contexts, they have only been established as a genuine field of historical study during the past two decades, owing much to Robert Rydell's continuous input.[6] Rydell argues that the expositions played a crucial role in the process of legitimating expansionist politics and racialized science in American society. The spectacular fairs, "upper-class creations initiated and controlled by locally or nationally prominent elites,"[7] were showrooms of a new mass culture and at the same time responses to the social chaos of the industrial era. Illusion was a central part of their setup. Pompous papier-mâché architecture in the European beaux arts style "represented efficient cities located in real cities marred by just the opposite: slums, corruption, and disease."[8] The fairs' amusement areas, often called "midways," served to still the public's appetite for entertainment while also channeling social unrest into harmless, escapist joy rides.[9] Besides merry-go-rounds, freak shows, and similar concessions, there were few profitable exhibits. The expositions were not a place of trade but a display case for industrial and agricultural goods. Therefore, long-term economic success had to be measured in terms of their symbolic impact.[10]

In the evolution of Rydell's work, world's fairs have increasingly been acknowledged as contested sites. While the elites tried to force their interpretations of progress upon a mass audience, resistance was ever present at the fairs. Some exhibits proved so unpopular that they had to be abandoned, and symbol-laden art displays were simply ignored. Women and African Americans fought over their portrayal at the fairs. As the last American exposition of the Victorian era, the PPIE deserves special attention. Thanks to the advancement of technology and visual media, it represented perhaps the most comprehensive attempt to illustrate the utopian belief in everlasting progress. At the same time, in the midst of World War I, a shadow of doubt was cast over its messages. In the light of this tension, it is surprising that the secondary literature on the San Francisco fair has so far remained modest in scope and depth.[11]

Bellamy's legacy had influenced the world's fairs ever since the exposition in Chicago in 1893. As "laboratories of modernization,"[12] the fairs celebrated the successful experiment of Western civilization. In the safe environment of artificial cities, middle-class spectators tasted the fruits of industrialization and imperialism. Displayed in entertaining exhibits, foreign people and mysterious machines lost their horrors. Journalists and writers, the same cultural custodians who had interpreted the building of the Panama Canal, supplied everyone who could not travel there in person with texts and images explaining the meanings of the expositions. "Fairs influenced popular belief systems, buttressing them with scientific, political, and religious authority."[13] And yet, despite their optimistic rhetoric and imagery, the expositions could not conceal the inherent contradictions of the new world they were showing and constructing.

These contradictions also resurfaced on another level. The expositions' role as national emblems should not conceal the fact that they were multidimensional by design. Not only were other countries represented as "competitors," which put the fairs in an international, even global, context similar to the Panama Canal project, but they were also expressions of local politics and social visions. The PPIE was about San Francisco, its past and future, and from this perception sprang most of the energy and enthusiasm invested into the fair.

After the Earthquake:
The Resurrection of San Francisco

For a world's fair celebrating the opening of the Panama Canal, San Francisco seemed a natural choice. Goods from the Philippines, the new American colony, went through the city's harbor, and it was ideally situated as a center for the expanding Pacific trade.[14] Business leaders had long held plans for an exposition. Roosevelt's intervention in Panama created a new scenario for their optimistic outlook. "The Panama Canal will probably be built," noted the public relations secretary of the local Chamber of Commerce in January 1904, the year of the St. Louis World's Fair.[15]

At the same time, reformers called for a complete overhaul of the city's layout. Advocates of the City Beautiful movement favored an urban architecture loosely based on Bellamy's Boston in the year 2000, expressing social harmony and wealth. In May 1904 the Association for the Improvement and Adornment of San Francisco asked the architect Daniel Burnham to design a master plan. As director of works, Burnham had been in charge of the outline and the buildings of the Columbian Exposition in Chicago. A few years prior to his engagement on the West Coast, he was involved in the remodeling of the national capital

Washington, D.C. The San Francisco upper class wanted him to give the city the look of a confident, expanding metropolis. In the fall of 1905, Burnham presented a plan with central boulevards, greenbelts, and imposing public buildings and squares.[16]

A few months later, shortly after Burnham's ideas had been publicly displayed, the city was devastated by an earthquake. Almost seven hundred inhabitants died, and three out of four residents lost their homes as a result of fires burning for three days after the convulsion on April 4, 1906.[17] The architect tried to argue for a total rebuilding of the city going even beyond his original scheme, but a tense political climate and long-lasting corruption trials, which would paralyze the city government for years, prevented the realization of his plans. Within half a decade after the catastrophe, San Francisco was restored in much the same way it had looked before.[18]

However, the project of a world's fair survived. In 1894 the city had already housed an exposition. Some of the exhibits were shipped directly from the Columbian Exposition in Chicago to Golden Gate Park.[19] But this time, an imitation of earlier fairs would not be good enough. San Francisco wanted its own grand vision of the future to make up for the losses of the past. However, the anticipated opening of the Panama Canal also inspired other cities to place a bid. In 1909 the much smaller town of San Diego, the southernmost port of California, entered into the race. A year later, New Orleans emerged as another competitor. San Diego officials finally agreed to forego the official recognition of their own fair in order to put San Francisco in a better position. With a "massive lobbying blitz of Congress,"[20] aided by promotional postcards, San Francisco beat New Orleans in a close decision. The Panama-California Exposition (PCE) in San Diego had a smaller budget and a regional focus. It was not seen as a rival event but as a supplement to the "big show"[21] in the North, and it counted 3.5 million admissions.

During the preceding decade, the local business elites had exerted a strong influence on San Francisco politics, even though the city was ruled by the Union Labor Party, which pursued an agenda based on the interests of organized laborers.[22] The corporate leaders viewed themselves as Progressives who strove to overcome partisan politics. After the world's fair decision in favor of their own city, the business community nominated James Rolph Jr., the owner of a shipping firm and other enterprises, as their candidate for the position of city mayor. In his campaign, Rolph, who had already lobbied for San Francisco as the world's fair host, stressed the redeeming role the exposition could play as a symbol of the city's rebuilding and civic unity. Rolph profited from the enthusiasm for the fair and beat mayor Patrick McCarthy of the Labor Party in the election on September 26, 1911. He remained in office until 1930.

On October 14, 1911, President Taft laid the foundation for the exposition in Golden Gate Park, one of the proposed fairgrounds at the time. Edward Bennett, who had worked with Burnham on his master plan prior to the earthquake, was responsible for the layout of the exhibition. Originally, by turning the entire city into a world's fair, a revival of the City Beautiful ideas seemed possible but then proved too expensive.[23] Eventually, the area east of the Presidio, at the time the largest army post in the United States, was chosen as the sole site for the PPIE. The exposition company evicted several hundred families from the area now known as the Marina District, and hundreds of acres of swamp land were filled in.[24] The only deliberate interaction with the rest of the city turned out to be the auditorium downtown, which hosted the fair's conventions. Afterward, it became part of the new Civic Center financed by the surplus of the exposition.

Another businessman, Charles C. Moore, president of a hydro-electrical engineering firm, headed the exposition company. The PPIE had a budget of $50 million. The city and the state of California provided $5 million each, in addition to the $7.5 million raised by selling stocks in the company to the public. Other states, foreign governments, and private exhibitors accounted for the remainder of the sum.[25] The fair received national recognition but no federal money. Its willingness to forgo a federal contribution had helped San Francisco to win the competition for the host city.[26]

West Coast ports such as San Francisco, Los Angeles, San Diego, Oakland, Portland, and Seattle hoped to profit from the economic benefits promised by the opening of the Panama Canal. To prepare for the increased trade, harbor facilities were overhauled and expanded, especially in Los Angeles. In contrast, the powerful railroad companies tried to downplay the significance of the seaway. Their managers feared that the transport of goods by way of Panama would prove less costly than by train in spite of the Canal fees.[27] Their efforts triggered little support from Progressive businessmen, though, who welcomed the new route as a means to diminish the influence of the anti-reformist railroads. "The canal has thus converted the Pacific Coast from a sparsely settled terminus for trickling lines of transcontinental traffic into a potential field for wide ramifications of commerce and industries radiating from western ports," the vice president of the San Francisco chamber of commerce declared after the Canal opening. "The Pacific Coast is thus entering upon a new era."[28]

It is important to point out the similarities between the interpretations of the Canal in the works of the Panama authors and by the organizers of the fair. In both cases, economic factors played a part but were overshadowed by the belief that the implications of the engineering feat would help defeat the social forces of the past and propel the whole region (or nation, respectively) into a new age of common ideals. In San Francisco, the world's fair became the place where this vision of the future, both for the nation and the city, would manifest itself. Like the Canal Zone, it was a laboratory for ideas that proved impossible to apply in the real world: "The virtues of the City Beautiful—political collaboration, class harmony, aesthetic unity—were acceptable only in the realm of fantasy."[29]

As in the Canal Zone with its discrimination against West Indian laborers, this collective vision did not embrace all parts of society. As Marie Bolton points out, the Progressive elite excluded the working poor in its plans for the revitalization of San Francisco. For many of them, the world's fair admittance fee of 50 cents proved too expensive. Bolton argues that the assertive policies of the city leaders to contain the poor were signs of underlying anxieties. Like the representatives of organized labor, they assumed that the Canal would bring floods of immigrants to the West Coast who would cause social unrest.[30] Roosevelt and other expansionists had used the building of the Panama Canal to construct a national identity neglecting women, African Americans, and immigrants who were not of Anglo-Saxon origin. In a similar way, the architecture and art of the PPIE spoke an exclusive language, celebrating the achievements of a white civilization. "The resumption of the advance toward the Orient"[31] compensated for the fear that the resurrection of San Francisco would prove a failure.

Inside the Walled City:
The Visual Language of the Exposition

The official poster for the PPIE showed the muscular, enlarged body of Hercules splitting the Panamanian rock. In the distance, the faint contour of the exposition skyline comes into view (fig. 22). The color

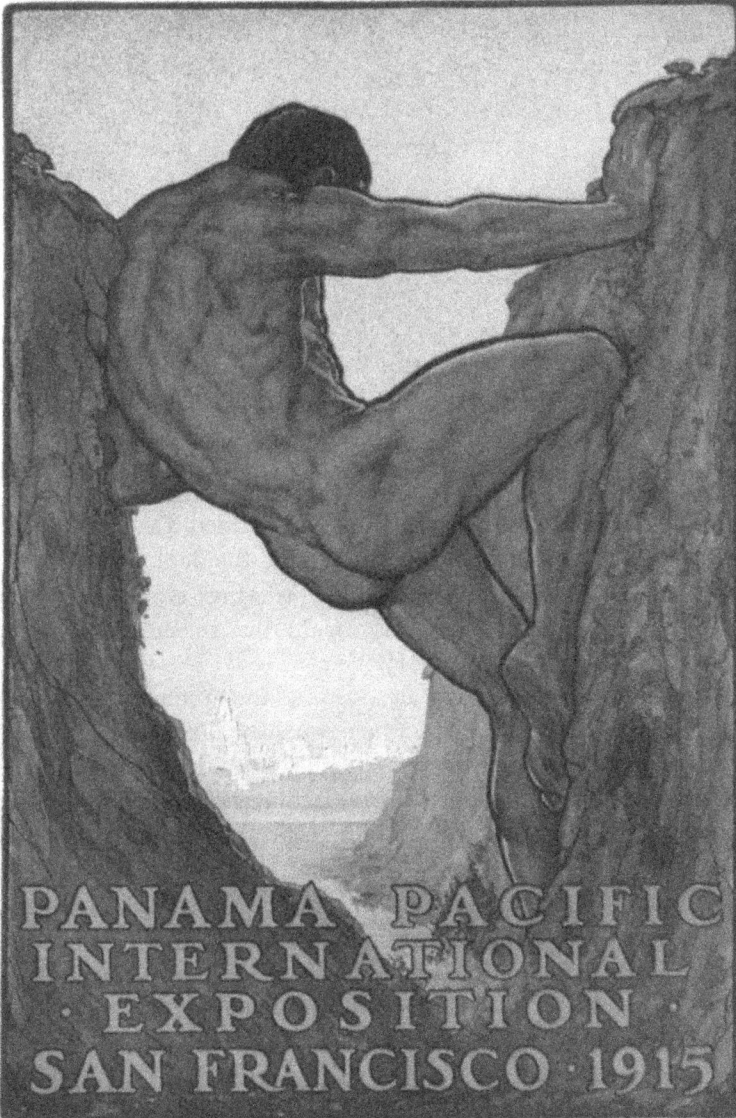

Figure 22. *The Thirteenth Labor of Hercules,* official poster of the World's Fair in San Francisco (reprinted with permission from San Francisco History Center, San Francisco Public Library).

Figure 23. Map of the fairgrounds in San Francisco (reprinted from *Official Guide of the Panama-Pacific International Exposition San Francisco 1915*, June ed. [San Francisco: Wahlgreen, 1915]).

lithograph by the artist Parham Nahl was entitled *The Thirteenth Labor of Hercules*. In the public poster competition, the depiction of an eagle cutting the Canal Zone with its beak, its wings embracing the two oceans, had come in second.[32] It is obvious that the eroticized figure of the superhero held greater visual power.

Greek and Roman models also inspired the architecture and sculpture of the PPIE, carried out in the popular beaux arts style employed at previous fairs. The main section of the exposition comprised eleven exhibition buildings arranged around courtyards and perpendicular avenues: the Walled City (fig. 23). The central axis ran from Chestnut Street—today the main thoroughfare of the Marina District—to the San Francisco Bay and featured the Tower of Jewels, the emblem of the PPIE, as well as two other signifiers, the Fountain of Energy and the Column of Progress, all of which were meant to represent the Panama Canal. The amusement concessions were located in the area east of the Walled City (called "The Zone"), whereas the pavilions of the different states and nations as well as the livestock exhibits spread to the west, bordering the Presidio.

The layout of the central courtyard, the Court of the Universe, and the Tower of Jewels was designed by McKim, Mead & White. The New York–based architectural firm had built the new Pennsylvania Station in Manhattan in 1910 (which has since been torn down) and, like Daniel

Burnham, played a crucial role in the remodeling of Washington, D.C. The firm's work had greatly contributed to the popularity of antique and renaissance architecture in the United States. The Court of the Universe was reminiscent of St. Peter's Square, the Column of Progress recalled Trajan's Column, and the gigantic Palace of Machinery resembled the Thermae of Caracalla, all of them located in Rome. The only prominent exception from this beaux arts frenzy was the California Building, characterized by the "mission style" typical of the region. In San Diego, the PCE organizers and their principal architect Bertram Goodhue had designed most of the fair architecture in a variant of the mission style, which was supposed to remind visitors of the Spanish heritage of the city.[33]

Coloring and lighting also played an important part in the design of the fair. The graphic artist Jules Guerin was named director of color and decoration, a first in the history of the world's fairs. In cooperation with Burnham's former assistant Bennett, he chose the basic color range that the architects were to employ. The most prominent color was an almost pinkish shade, a "faint ivory"[34] used in the imitation travertine of the building façades, not unlike the tone dominating the fair in St. Louis. (Chicago had literally been a white exposition.) The color scheme gave the PPIE an almost Mediterranean appearance, which was enhanced by the geography of the city "between the tawny Grecian hills and the blue Italian seas which are California's," as one contemporary observer noted.[35]

At night, special light effects illuminated the buildings from varying angles. These were overseen by William D'Arcy Ryan, director of the lighting lab at General Electric. Floodlights were mounted on ships anchored in the San Francisco Bay. The Tower of Jewels took its name from the fact that thousands of jewels, "each backed with a tiny mirror,"[36] reflected the incoming light (fig. 24). Ryan went even as far as to create artificial sunsets, thus giving the World's Fair "the effect of some dream city of the Arabian nights."[37]

The visual spectacle of the PPIE reflected not only the technological achievements of the past decade but also the complexity of the overall message the organizers had in mind. The whole fair was a work of art, of carefully constructed meanings, intensified by "the beauty of design and of plastic form, the interest of varying vistas and varying shadows, the satisfying sense of proportion and the thrill of perspective."[38] As Paul Greenhalgh points out, the classical architecture of the world's

Figure 24. The Tower of Jewels illuminated at night (reprinted from *The Blue Book, a Comprehensive Official Souvenir View Book of the Panama-Pacific International Exposition at San Francisco 1915* [San Francisco: Robert A. Reid, 1915], 19).

Figure 25. The Palace of Fine Arts (photograph by the author).

fairs served to suggest "solidity and permanence,"[39] qualities that were in high demand after the San Francisco earthquake. At the same time, it was all show: the buildings were to be torn down after the exposition closed, and almost everything was made out of imitation materials and reinforced with plaster or concrete, even the statues.

There was one exception to the general impression of the Walled City: Bernard Maybeck's Palace of Fine Arts (fig. 25). The building was arranged in a semi-circle around a lagoon and housed 120 art galleries. Most of them presented American works, but among the exhibits were works by the Italian Futurists, forerunners of modern art and shown for the first time in the United States. Although the Berkeley architect adhered to the classical theme of the exposition, designing a colonnade and a rotunda, the palace conveyed a different atmosphere than the rest of the exposition buildings. "He wanted to suggest a splendid ruin, suddenly come upon by travelers, after a long journey in a desert. He has invested the whole place with an atmosphere of tragedy," wrote a contemporary author.[40] Ironically, the building proved so popular that it was not demolished, and today it remains the only actual remnant of the fair. It was rebuilt in steel and concrete in the early 1960s and now houses a science museum. Maybeck explained that he had tried to design a building that would match the seriousness of the paintings it displayed, "sadness modified by the feeling that beauty has a soothing influence."[41] But perhaps the popularity of his artificial ruins also implied the failure of the main architectural program to reflect the spirit of the recently devastated city.

The exposition was officially opened on February 20, 1915, by remote control: President Wilson turned a golden key in his office in the White House. "The electric current carried the message to San Francisco, an instant later releasing the waters of the Fountain of Energy."[42] The fairgrounds remained open for 288 days. The exposition company made sure that visitors and journalists spread the word (and the image) throughout the nation. It awarded special concessions for the official publications, guidebooks, and photographs to three different firms. The official photographer was the Cardinal-Vincent Company, which had about a hundred employees including salesmen. It was also granted the postcard concession.[43] In addition, countless booklets and articles on every aspect of the PPIE were published by independent authors. Just like the building of the Panama Canal, the fair was captured and, most importantly, explained for a national audience.

Interpretations were necessary even for the visitors who came to San Francisco and looked at the buildings and exhibits with their own eyes. The aesthetic program of the fair was crammed with allegories and metaphors. In a crude translation from engineering work into beaux arts architecture, the Panama Canal was referred to by the structures on the central axis, the Tower of Jewels, the Fountain of Energy, and the Column of Progress. The tower, designed by New York architect Thomas Hastings, rose 435 feet high but had originally been planned 100 feet higher.[44] "This great tower symbolizes the Panama Canal, the jewel today that is most resplendent," one critic wrote.[45] It was flanked by equestrian statues of the explorers Cortez and Pizarro. Sculptures on the tower depicted the four figures of an adventurer, priest, philosopher, and soldier—embodiments of the action and spirit that characterized Western civilization. Courtyards on either side featured two small fountains with utopian themes, the Fountain of Youth and the Fountain of El Dorado. The tower's arch held two monumental murals on the Canal building, which are discussed at the end of this chapter.

The sculptural program of the PPIE was overseen and partly executed by the Austrian Carl Bitter, who had been director of sculpture at the world's fairs in Buffalo and St. Louis, and A. Stirling Calder from New York. The Fountain of Energy (fig. 26), whose waters had been set in motion by President Wilson's key, was Calder's main piece. Viewed from the entrance, it stood in front of the Tower of Jewels and, not surprisingly, was also supposed to represent "the triumph of the Panama Canal."[46] In the center of the fountain, an immense globe held by mermaid sculptures and encircled by a band—alluding to the completed waterway—arose. The figures of a woman and a man with heads in the shapes of a cat and a bull stretched across the sphere, representing Eastern and Western civilization. On top was the horseback figure of "Energy; the lord of the Isthmian way"[47] crowned with outstretched arms, a scene referred to by a poet as "The Coming of the Superman."[48] On his shoulders rested two allegorical figures presumably representing Fame and Valor. Like other allegories in the fair's sculpture, they prompted frequent criticism. One author thought of them as "utterly irrelevant and unnecessary."[49]

The 160-feet-high Column of Progress (fig. 27) stood at the northern end of the Court of the Universe, which bordered the Tower of Jewels in the direction of the San Francisco Bay. Again, the figure of a man, sculpted by the artist Hermon A. McNeil, was placed at the top of a

Figure 26. The Fountain of Energy (reprinted from *The Blue Book*, 25).

Figure 27. The Column of Progress (reprinted from *The Blue Book*, 268).

Figure 28. The sculpture *The End of the Trail* (reprinted from *The Blue Book*, 48).

frieze. This "Adventurous Bowman," a representation of "the leader, the achiever, the man who dreams and dares,"[50] pointed his arrow toward the ocean. The kneeling figure of a woman presented a laurel wreath and the palm of victory to him, "the reward of his glory and the encouragement of her hope."[51] The critic John Daniel Barry complained that McNeil's depiction was "distinctly old-fashioned. He made the archer a superman, pushing forward by force, and by the dominance of personality." Instead, Barry would have preferred the artist "to express the new spirit of today, the spirit that honors the common man and that makes an ideal of social co-operation."[52]

The most popular sculpture at the PPIE turned out to be an artistic work that was symbolic in a much more naturalistic sense: James Earle Fraser's *The End of the Trail* (fig. 28). It portrayed a fatigued Native American resting on his horse, exhausted by his own and his people's struggle for survival.[53] Compared to the sights discussed above, it was placed in a rather unassuming location in front of the small Court of Palms. Fraser's Native American was interpreted by a critic as the portrayal of a "racial tragedy."[54] The sculpture was often contrasted with another equestrian statue, *The Pioneer* by Solon Borglum, even though they could not be looked at simultaneously. Borglum's pioneer, "very typical of the white man and the victorious march of his civilization,"[55] stood near the Court of Flowers in another section of the fairgrounds. Underneath the superficial message of the white man's triumph over the stubborn Indian, *The End of the Trail* conveyed a sense of loss, which seemed to appeal to the fair's visitors ("It is admired by all kinds of people"[56]) in a similar way as Maybeck's Palace of Fine Arts.

The "racial tragedy" of the Native American on horseback also pointed to a common theme in the sculptural work at the PPIE and earlier fairs: the triumph of Western man, here epitomized by the completion of the Panama Canal, after an evolutionary struggle between the races—and even the sexes. As in the other world's fairs, these social Darwinist interpretations resounded in the anthropological exhibits and in the Native villages of the fair's amusement zone. In the sculpture, they reached almost absurd levels. The work *Mutation,* for instance, by Chester A. Beach, depicted two symbolic figures in the Court of Abundance. "They show a man and a woman in the throes of the struggle from lower to higher planes. The man's old animal self can be vaguely seen, a crude hand, gripping his foot and trying in vain to hold him back. The woman struggles upward out of the clutch of intellectual slavery,—a veiled figure, barely distinguishable at her feet."[57]

In the murals on the walls of the Tower of Jewels, the use of allegorical figures, modern elements, and social Darwinist thought all came together in an unusual blend of symbolism. Due to the limited color scheme, the artists were restricted to a palette of five shades. The main works of painting were placed on either wall under the arch of the Tower. These gigantic triptychs, two hundred feet long and sixteen feet high, illustrated the achievement of the Panama Canal and were created by William de Leftwich Dodge. On the right, the central panel entitled *Gateway of All Nations* (fig. 29) was flanked by *Labor Crowned* and *Achievement.* In *Gateway,* the nude male representation of Purpose makes his way through the Canal, encouraged by the winged figure of Fame, sounding her trumpet. "One figure, with covered face, flees from the appeal of the siren, but whom he represents, or why he flees, I cannot tell," admits the critic Ben Macomber.[58] On the left side of the panel, in a notable depiction of technology, a steam shovel is shown. "Beside it the

Figure 29. The mural *Gateway of All Nations* (reprinted from *The Blue Book*, 292).

Figure 30. The mural *Atlantic and Pacific* (reprinted from *The Blue Book*, 293).

master-workman whose brain has dominated all this, stern, alert, un-
tiring. At his feet, his labor done, crouches the man who has worked
merely with his hands and who now sits inert with fatigue," guide book
writer Katherine Burke suggested as an explanation.[59] The panels on
either side each show a muscular male figure in the center (denoting
Achievement and Labor, respectively) flanked by laborers coming for
their rewards. In *Labor Crowned*, a soldier heads the group, "thus acknowledging that the American Army led and directed this great work
of peace."[60] In *Achievement*, a grateful Panamanian "acknowledges the
benefits bestowed by civilization."[61]

On the left wall, the main panel was called *Atlantic and Pacific* (fig. 30),
with *Discovery* and *The Purchase* next to it. In the center of *Atlantic and
Pacific*, the figure of Labor joins the hands of the two oceans. To the right
side of the scene, there is a group of people from different nations who
supposedly all benefit from the Canal. To the left, a group of pioneers—
"who crossed the Isthmus and the plains in earlier days,"[62] but who were
also interpreted as allegories of the miner, prospector, engineer, and
farmer[63]—gathers around a prairie wagon. Before them, an Indian
sinks into the Pacific. On the smaller panel *Discovery*, Balboa, discoverer
of the Pacific Ocean, is shown with his followers, next to an Aztec Indian who "sits in the ruins of his home."[64] On the panel *The Purchase*,
beret-wearing Frenchmen lie down their tools, while an allegory of
France hands America (who is accompanied by Ambition) a scroll, presumably with a title of possession for the Canal.

The murals were reminiscent of the way European nation-states celebrated their military victories in former centuries. But instead of weav-
ing a tight net of meanings that would have been obvious to contemporary observers and thus hard to escape from—as in, say, Louis XIV's
palaces at Versailles—the allegorical representations frequently confused the critics of the PPIE. They were also contrasted with naturalistic depictions of nude bodies and machinery that had little in common
with classical painting. The murals were blatantly racist, likening the
Canal building to the defeat of "primitive" populations and the overciv-
ilized French. At the same time, the taking of the Canal Zone was viewed
as an extension of manifest destiny comparable to the endeavors of early
American pioneers.

The Walled City, the interpretive heart of the PPIE, "pointed back-
wards with sentiment to the ancient world and revived American frontier rhetoric through architecture and sculpture."[65] Into the language

of dominance and exclusion, the theme of universal friendship—
expressed by the joining of the two oceans in Dodge's left triptych and
complementary sculptures such as the *Nations of the East* and the *Nations
of the West* on top of the arches flanking the Court of the Universe—was
incorporated without any apparent sense of contradiction. What was
the message of this eclecticism? "The inspiring lesson of beauty, expressed so simply and intelligently, will sink deep into the minds of the
great masses, to be reborn in an endless stream of aesthetic expression
in the spiritual and physical improvement of the people," the critic
Eugen Neuhaus predicted.[66] Others, as noted throughout this chapter,
attacked the complexity of the fair's art: "Symbolism is here carried to
an extreme that spoils the simplicity which alone makes a really great
work imposing."[67] The cultural historian Jackson Lears argues convincingly that visual art assembled from "the grab bag of symbolic forms"
(such as the post–Civil War "revivals" in architecture) trivialized its
transformative power and made Americans believe that their urban
environment was "somehow artificial and unreal"[68]—which, of course,
was literally true for the Walled City. It is little surprising then that the
most successful pieces of architecture and sculpture at the PPIE expressed a sense of loss in the post-earthquake San Francisco.

The symbols of the Panama Canal—Hercules, Energy, the Adventurous Bowman—were supermen who reflected Theodore Roosevelt's
successful attempt to present the Canal project as a triumph of American manliness. Although some of the sculptors were female, women
played only a minor role in the depictions at the fair and were reduced
to the role of allegories or admiring sidekicks.[69] Not all contemporary
commentators shared this view, though. While John Daniel Barry complained that the exposition did not reflect "the new social and economic
ideas and the changing relations between women and men,"[70] Stella
Perry wrote in a women's magazine that both sexes were represented
equally—even in the sculpture—and applauded the fact that there were
no special exhibits for women.[71]

Exhibiting the Panama Canal: Displays and Concessions

While the architecture and art of the San Francisco World's Fair found
its inspiration in the past, the exhibits presented a mind-boggling picture

of technology's transformative power. Behind the fake travertine façades, the newest machinery from all areas of life was on display. New ideas were pervading not only manufacturing, transportation, and agriculture, but liberal arts, education, and social economy as well. Each of these keywords received its own exposition palace in the midst of the Walled City. With a few exceptions, nothing "was intended to be shown which had not been discovered or invented, or the process or application of which had not been substantially developed, within the last decade." The exhibits were to be "the latest words, the newest thought, the edge of day for the new world."[72]

Every day, eighteen Ford Model T cars were produced on site in the Palace of Transportation. The world's largest typewriter, by the news syndicate Underwood & Underwood, had superhuman dimensions. Alexander Graham Bell, the inventor of the telephone, called from New York and spoke to the same Thomas A. Watson to whom he had placed the first phone call in history almost sixty-one years before.[73]

Movies, another novelty, "gave the Panama-Pacific International Exposition a singular advantage over its predecessors." There were films, "stereomotorgraphs," photographs, photocopies, models—"every conceivable sort of visual representation."[74] Foreign places were brought to the fair in pictures—including the Panama Canal, the subject of a government film shown in the Palace of Mines.[75] The exposition provided movies in dozens of different locations and was itself advertised by promotional films shown in 3,500 theaters across the country. Film director D. W. Griffith contemplated producing a movie on the fair that would rival *The Birth of a Nation*.[76] Franklin K. Lane, secretary of the interior, said of the PPIE: "It is a great moving picture."[77]

To the west of the Walled City, beyond the Palace of Fine Arts, the visitor was surrounded by the buildings of the nations and states represented at the fair. Due to the outbreak of World War I, some countries had pulled back, and in the end only twenty-two foreign governments operated their own buildings, while nine others supplied exhibits shown elsewhere on the fairgrounds. The encounter with Cuba, Panama, and the Philippines was meant to remind the fairgoer of U.S. colonial ventures. The coincidence that Panama was also planning an exposition on its own territory, which consumed its organizational energies, resulted in the curious fact that the Panamanian pavilion "remained empty and closed throughout the season."[78] Consequently, the public perception of the country was completely dominated by the construction of the

Panama Canal, which the PPIE organizers, like the Panama authors, viewed as a *domestic* event. The popular Philippine exhibits, on the other hand, were part of another discourse, serving to explain the blessings of colonialism for the conquered people.

Confronted with "a panorama of a Nation's idealism and a people's prompt response,"[79] the visitor came across the Philippine Islands not only in their own building, but also in the Palace of Education and Social Economy as well as in the Palace of Agriculture. The national pavilion displayed handcrafted products, agricultural goods, and picture galleries with ethnological motives, images of public works and other photographs, "showing what opportunities the islands held forth to the settler, the investor, and the tourist."[80] The orchid conservatory with more than two thousand plants proved especially popular. The Palace of Education showed the progress made in schools and hospitals and was symbolic of the U.S. "intention to help and uplift rather than merely to exploit."[81] Due to their success, parts of the exhibits were later shipped to the PCE in San Diego.[82]

Live displays of people—highly inappropriate by today's standards— were a common feature of the world's fairs. The Italian educator Maria Montessori had thirty-five children in a glass pavilion exhibited to demonstrate "self-education." In the Zone, the infant incubators held actual babies. At the Race Betterment booth in the Palace of Education, set up by a movement aiming "to create a new and superior race through personal and public hygiene," four people "sat in vibrating chairs and were agitated physically by electric motor, and usually looked resentful of the past and careless of the future, and as though they thoroughly needed the good shaking they were getting."[83]

The Race Betterment Foundation, headed by A. J. Read, also displayed "large plaster casts of Atlas, Venus and of Apollo, Belvedere type, to advertise the human race at its best,"[84] and its display fit well within the social Darwinist themes of the sculpture and murals in the Walled City. In relation to earlier world's fairs, the focus had shifted from simple anthropological exhibits—which were still a big feature at the PCE in San Diego—to the "scientific" promotion of race hygiene and eugenics. The stereotypical exposure of foreign people also took place in the Zone but came increasingly under pressure from protesters. The Hawaiian Village had to be renamed as Hula Dancers to imply a more folkloristic purpose, and the opium den in Underground China was closed down. "Native villages did not seem to do well," noted the official historian Frank Morton Todd, one of the few people who had anything to say about these exhibits. Apparently, the novelty effect had already worn off. Since a group of Somalians "failed to commit any acts of cannibalism," they turned out to be an economic failure. When they refused to leave the premises, a platoon of guards escorted them to the Yacht Harbor, from which they were deported.[85]

The PPIE was also the host of 948 meetings (or congresses) organized by professional associations. Among other topics, they discussed implications of the Canal building, such as with regard to sanitation.[86] Former president and historian Theodore Roosevelt, by now a political outsider, spoke at the Panama-Pacific Historical Congress. He was enthusiastically greeted "with shouts and cheers and flags and flowers."[87] A great crowd gathered in the Court of the Universe, and TR enjoyed a rock star–like appearance. "Who built the Panama Canal?" PPIE President Moore asked in his introduction. "Teddy!" yelled the audience.[88] Canal builder Goethals, now a major general, served as honorary president of the International Engineering Congress.

The main amusement avenue located to the east of the main exposition buildings had been named "The Zone" with respect to the Panama Canal Zone. Out of thirty thousand submissions in a prize competition, six people had come up with this term. The award, a $10 season book to the fair, was assigned to Mrs. J. Cortissoz of San Francisco, whose letter was the first to be opened.[89] In San Diego the amusement mile was called "Isthmus." Whereas the country Panama was present only as an empty building at the PPIE, as an absence of meaning, the Canal Zone resurfaced in the shape of an amusement zone, a site for all kinds of projections and purposes.

The midway, often referred to as the "Joy Zone," was a great contrast to the architecture and sculpture next door. "There is no harmony whatever in the Zone anywhere, either in the form, style, or color, unless it be the harmony of ugliness which is carried through this riotous mêlée of flimsiness and sham," art critic Neuhaus complained.[90] But such Victorian condemnation was rare; commentator Edith Stellman, on the other hand, viewed the infantile entertainment as "a sort of progressive circus," an education for adults: "On the Zone we are all children, regardless of size or years."[91] Midway veteran Frederic Thompson had designed Toyland Grown-Up, the largest concession at the fair. Visitors entered a strange world featuring fantastic architecture and outsized toys that were supposed to make people feel small again. Automatons took over stressful human tasks, and labor settings were presented as playgrounds. Thompson, like Bellamy before him, suggested that work in a modern society would assume the role of play.[92]

Quite often, the Zone was discussed as an economic disappointment. It seemed difficult to predict what people would like. *Stella,* the simple painting of a nude woman that had already been exhibited in other places, became "the hit of the place." Visitors streamed into a little theater with no seats where they could not stay long. With an admission fee of 10 cents, *Stella* took in more than $75,000, even though, as Todd noted, nudes by famous painters could be seen for free at the Palace of Fine Arts.[93] Other venues, especially the well-known types of attractions, did not make any money. The director of concessions and admissions called the Zone "a partial success."[94] Many reasons were given, such as the large number of offerings and the lack of innovations.

One of the most successful concessions was the Panama Canal exhibit, conceived by a private company and located off the main avenue on Goethals Street. The federal government did not have a building of its own on the fairgrounds and instead spread its various exhibits on topics such as public education, health, science, and agriculture throughout the different halls. The Panama Canal was also the subject of an exhibit under the auspices of the Government Printing Office in the Palace of Liberal Arts. It displayed a small topographical model of the Canal as well as a replication of a Canal Zone hospital and appliances for exterminating mosquitoes. A complete set of the *Canal Record* and the annual reports was available for reading.[95] Compared to the concession on the Zone, however, these exhibits were of little significance.

The Panama Canal concession had been the idea of the Chicago engineer L. E. Myers "while on a visit to Panama in 1911."[96] It took his company, a business specializing in the building and operation of public utilities, two years to prepare the model of the seaway and the surrounding countryside.[97] The result was a technological construction of staggering dimensions—an amphitheater with a spectators' gallery revolving around a detailed miniature of the Canal Zone: "At an expenditure of over $500,000 the original canal and the surrounding zone territory have been reproduced in miniature form so accurately one can almost imagine that he is taking an aeroplane trip over the Isthmus of Panama. A birdseye view of the entire country is obtained as the moving platform slowly conveys one over the five-acre tract of land upon which has been constructed this clever piece of engineering work."[98]

In the planning stage, the government did not approve of the undertaking, which clearly dwarfed its own Canal model. "Through the Exposition Officials, the Government objected to the reproduction of the Panama Canal by private interests and all work was temporarily halted."[99] But since none of its own concepts could rival Myers's ingenuity, the government eventually abandoned its opposition and instead decided to give the concession an official blessing. "A complete, correct and faithful working reproduction of the Panama Canal and Canal Zone has been constructed," Major Boggs from the Washington office of the Canal Commission confirmed in a letter to the Exhibition Company. He concluded that a visit to the Zone "will in half an hour impart to anyone a more complete knowledge of the Canal than would a visit of several days to the waterway itself."[100] Just like the original, the model was hailed as a triumph of engineering.

The Canal model was operated by the prominent showman Fred W. McClellan and became the greatest box-office success on the Zone. With an admission charge of 50 cents, it grossed more than $338,000.[101] Souvenirs of the Canal, among them paperweights filled with soil from Culebra Cut, could be bought at a booth nearby.[102] Through an impressive beaux arts building with columns (fig. 31), which resembled the architecture in the Walled City, the visitor entered a sheltered amphitheater with a capacity of 1,200 seats. These "opera chairs"[103] were attached to a moving platform that took twenty-three minutes to revolve around the model of the Canal Zone located twenty feet below the spectators, "the effect being that of a general bird's-eye view" (figs. 32 and 33).[104] On the vertical walls, paintings by artists created the illusion of a horizon. Commentators recited the enormous amounts of material that had gone into the production of the exhibit: 2 million feet of lumber, 217 tons of cement and plaster, and 85 miles of copper wire.[105]

The copper wire was used in a mechanism that may seem unremarkable today but that was nothing short of revolutionary at the time. While seated, each visitor held a "duplex telephone receiver," the equivalent of modern headphones, and listened to a continuous lecture explaining what could be seen below. Each time the platform moved, a new part of the lecture became available for listening. There were fifteen different lecture segments, each recorded on three individual "phonographs" used for the operation of the exhibit. A set of fifteen additional records was reserved "for emergency use."[106] The encounter with the Panama Canal thus became an experience of two senses. "The lecture was so timed that the auditor had an opportunity to carefully view the objects described. The concluding sentence of the description of the canal was, 'Do not forget your packages.'"[107] The manuscript had been written with the help of Canal Zone guide William A. Baxter Jr. The lab of the famous inventor Thomas A. Edison contributed its expertise to the technological rendering.

The visitors were thrilled. Edith Stellman described her encounter with a friend, a college professor, in the Zone: "'Come on,' he cried, waving a bag of popcorn at me. 'This is great! Have you ridden around the Panama Canal?' I endeavored to reply, but he did not wait for me.

Figure 31 *(above)*. The Panama Canal building in the fair's "Zone" (reprinted with permission from San Francisco History Center, San Francisco Public Library).

Figure 32 *(top right)*. The Panama Canal exhibit (reprinted with permission from San Francisco History Center, San Francisco Public Library).

Figure 33 *(bottom right)*. The spectators' gallery at the Panama Canal exhibit (reprinted with permission from San Francisco History Center, San Francisco Public Library).

"PANAMA CANAL BLDG." DEC. 31, 1914
EXPOSITION GROUNDS, SAN FRANCISCO 1915
THE L.E. MYERS CO. CHICAGO.

Figure 34. Miniatures at the Panama Canal exhibit (reprinted from *The Blue Book*, 309).

'It's the Canal itself to the life. Just as though you took the real trip,' he gasped."[108]

Like the books and articles of the Panama authors, the model provided interpretations of remote events; it served as a production site for second-order observations and meanings. While Roosevelt's photograph atop the steam shovel was a metaphor for the American effort on the Isthmus and, physically, a "flat" image on a newspaper page, the Panama Canal exhibit was a visual and sensual experience in itself and held more than just symbolic power. It was, in the words of its promoters, "the largest reproduction of any subject ever created,"[109] the perfect miniature. The original plan had been to allow visitors to get on boats and travel through the Canal Zone, which apparently proved too complicated.[110] In the end, electric engines and magnets pulled the model ships and trains (fig. 34). The impact on the spectator was predictable: the visitors viewed the Canal Zone exactly as Bishop and the photographers had captured it, from the panoramic perspective.[111] "When first you take a seat on the platform, you are looking simply at a vast colored model of the Panama Canal Zone, with its miniature mountains, rivers, locks, lighthouses, steamers, wireless telegraph towers in operation and distant vistas. But as you look longer and longer, the mountains seem to rise, the colors of the panoramas give the effect of mists, the distances become increased, parts of the map hundreds of feet away seem hundreds of miles. You watch the tiny craft and locomotives as one gazes from the top of a mountain. You feel that you are really looking at the canal itself. You leave the great enclosure with a little gasp of wonder and surprise."[112]

In contrast to the art and architecture of the Walled City, the Panama Canal exhibit managed to transmit a strong, unified audiovisual message to its audience and establish what Bill Brown calls "the proper power relation between spectator and spectacle, with the commodified image mimicking the commodified territorial possession that the Canal Zone itself had become."[113] The power of the exhibit clearly benefited from the "super-vision" that the government and the Panama authors had generated earlier through their photography and other visual representations of the Canal.

Michel Foucault used the concept of the *panopticon* to describe the discursive nature of observation. The central watchtower of a metaphorical, transparent prison signifies the omnipresence of surveillance, making specific acts of observing superfluous, deconstructing the relationship between overseer and prisoner. Anne McClintock applied the idea of the panoptical principle and its application as commodity spectacle to the Great Exhibition under the glass roof of the Crystal Palace, at the first world's fair held in London in 1851, with "its ability to merge the pleasure principle with the discipline of the spectacle."[114] The Panama Canal model went even one step further: As the spectators' gallery moved around effortlessly, the exhibit transformed the task of observing the American empire into a "remote-controlled," passive, and yet social act of consumption. Implementing this commodified, globalized gaze, the Canal miniature foreshadowed the cultural framework of our own age. At the fair, the exhibit, the "great enclosure," became a manifestation of not only the triumph on the Isthmus but also of the American resolution to create a new society, literally a *model* for the future.

"United, Co-operative, Productive": Fair Evaluations

In his discussion of the exposition poster, Brown concludes that the figure of Hercules "occults the mechanical achievement by refiguring the decade-long canal construction as the gesture of the individual, who is *whole* without being *part* of a labor, technological, or military force, without being a 'tool of the government.'"[115] Does this analysis, which contradicts most interpretations of the Panama authors, actually hold true with respect to the fair? Not if the poster is regarded within the context of the exposition. Writing on the occasion of the opening of the PPIE, Secretary of the Interior Franklin K. Lane praised the Panama Canal as the triumph of a cooperative society: "In three generations we have marched across a continent wider than Europe and crowned our achievement at our westernmost door with an exhibition of the worthiest products of our civilization. This we can say proudly is what a democracy can do. We are coming to a fuller national consciousness not merely as a nation among the family of nations, but as a people who have common interests and can collectively do things for themselves which it would be too great a hazard to leave in private hands. The building of the Panama Canal is a long step in the making of this nation, for it has given us pride in our ability as a modern working machine."[116]

While Lane repeated Bishop's metaphor of the state as a "working machine" from the federal government's point of view, the progressive San Francisco businessmen had their own reasons to promote the Canal and its fair as emblems of a united effort. The PPIE was their instrument to overcome partisan politics and celebrate the resurrection of San Francisco after the earthquake had destroyed not only the existing city but also the more daring plans for a City Beautiful. Years before the fair took actual shape, exposition company president Moore pitched its "progressiveness" by arguing that the exposition commemorated a contemporaneous event and not a historic one — unlike the St. Louis World's Fair in 1904, which celebrated the centennial of the Louisiana Purchase: "The great world's fairs preceding ours have been anniversary celebrations: complacent occasions, certainly, giving cause for congratulations, for reunion, for pledges for the future. Ours is a nuptial feast, if you please: a joyous send off, the auspicious completion of preliminaries and the beginning of a new life, united, co-operative, productive; a glad bridal celebration, at the home of the bride."[117]

Hercules was a fitting symbol for the interpretations of the Panama Canal. He embodied the manliness of the project that Roosevelt had preached, "the technologized and scientized"[118] aspects, and, last but not least, the racialized body of Bellamy's utopian state. He is a universal figure representing the collective (or public) rather than the individual (or private) effort needed to build the waterway and realize the fair. Like the architecture, sculpture, and paintings at the exposition, the poster also aestheticized, even eroticized, the engineering work. Anne Maxwell suggests that the PPIE managers had sensed a public wariness of overseas expansion and therefore avoided a more bellicose image.[119] But the artists at the fair did not hesitate to praise the achievements of Western civilization in blatant, aggressive ways. Although their depictions may have proven ineffective and confusing, the modern representations of the Canal, like the Panama Canal exhibit, managed to convey a better sense of progress and expansion as the American ways of life.

Like the Canal promoters, the makers and interpreters of the exposition combined nationalistic rhetoric with calls for "world's" progress, peace, and international understanding. While the seaway, Whitman's passage to India, was the physical symbol of a network between nations, the PPIE, like all world's fairs, displayed the results of international cooperation as well as competition. "Never before has it been brought home to the minds and hearts of men that the whole world is only one great country and that the goal of the highest and best type of internationalism is a 'United States of the World,'" wrote a college president in the booklet *The Legacy of the Exposition,* a collection of letters to the PPIE management published in 1916.[120] Many of the commentators applauded the fair's contribution to the friendship between nations. The worldview of the San Francisco businessmen resonates in their evaluation of the Panama Canal and its fair: "Trade, that greatest of all pacificators, promises to bring about that closer relationship between all peoples that shall ultimately do away with struggles between the races for trade and territory, for with the breaking of the barriers of distance and racial hatred the nations come into a co-operative basis in reference to their trade and commerce."[121] While the keyword "co-operative" recalled the collectivism of the Panama authors and their ideal Canal Zone state, the businessmen's "pacificator" trade moved goods across the Pacific Ocean, turning San Francisco and its fair into utopian symbols of the modern world and the coming of a Pacific Age.

At first, the outbreak of war in Europe seemed to support these interpretations. David Starr Jordan, the chancellor of Stanford University, asserted: "[T]he cure for war is the extension of patriotism. The Exposition stands for 'planetary patriotism.'"[122] Still, the exposition managers worried that the conflict would have a negative effect on the attendance.[123] Slow business in the amusement zone has also been explained as a reaction to the war.[124] On the other hand, American tourists who had considered a trip to Europe may have chosen to attend the PPIE instead. The fair, like its predecessors, was a place to *escape* reality and catch a glimpse of a promising future.

But it was precisely this vision of everlasting progress, salvation by technology, and conquest without battle that the war in Europe slowly began to undermine. Prior to the American involvement in the conflict, observers pointed out the war metaphors employed in the fair's social Darwinist art and exhibitions and in the building of the Panama Canal: "Each event is in its way an unexampled exhibition of war. One is a war between different races of mankind. The other is a war waged by the whole human race for the subjugation of the forces of nature."[125] As time passed, the war confirmed the underlying fears of failure and destruction embedded in the perception of San Francisco, its fair, and the nation as a whole. The door of escape into the future seemed to close. "Instead of exulting in his achievements, man had to mourn the perversion of industry, the degradation of life, the turning of productive into destructive agencies, the devotion of science to wholesale woe instead of healing, the waste of accumulated wealth on a scale only to be measured by many times the whole amount of many in the world," Frank Todd wrote in his postwar history of the exposition.[126]

Summing Up: The Final Chapter of Canal Interpretation

The PPIE brought the reality of the Panama Canal closer to the United States. Books, magazines, and photos had explained the waterway to the American middle class and reiterated the same interpretations with regard to the fair and its celebration of the Canal. In addition, a few million visitors experienced the exposition in person, looked at artworks such as the Fountain of Energy, or took a ride in one of the "opera chairs" at the Panama Canal exhibit. The cultural work of celebrating a historic event and constructing a vision of progress had never been more complex and comprehensive than at the fair in San Francisco. Color schemes, lighting effects, motion pictures, miniatures, and all other kinds of visual media created an all-encompassing illusion of reality.

The lessons of the Canal project were shown and praised at the fair. On the exposition poster, the figure of Hercules splitting the

Panamanian rock symbolized not only the triumph of American manliness and white civilization but also the superiority of a collective endeavor carried out by the government over private interests. All of these implications were projected onto the fair as an epochal achievement in itself. Even though the PPIE was planned by private businessmen (and the Panama Canal model constructed by a private company), the rhetoric of these city leaders differed little from the visions of Theodore Roosevelt and the Panama authors. The fair, like the Canal, was an instrument to overcome the corruption and partisan struggles of the past and enter an age of cooperative trade, civil unity, moral and technological progress.

The Panama Canal and the Panama-Pacific International Exposition were both interpreted in seemingly contradictory ways: as expressions of the past and the future, of world friendship, and of the struggle between the races. While the Canal Zone was described as the "birthplace of American history" as well as the model for a future society, a similar dichotomy was presented to the visitor at the PPIE. Like Hercules on the exposition poster, the neoclassical art and architecture of the Walled City was based on traditional European designs, and only a few hundred yards away, the rotating Panama Canal exhibit or the Model T assembly line demonstrated the latest technical innovations. The Canal and the PPIE were both hailed as agents of universal peace, and at the same time they were used to reinforce racial and sexual hierarchies and legitimate American expansion abroad.

California, the westernmost state and gateway to the Pacific Ocean, was a fitting location to celebrate national progress. On the other hand, the PPIE played a crucial role in a local context, the history of San Francisco. After the earthquake and the failure to reconstruct the metropolis as a City Beautiful, the fair became the main focus of the city leaders' visions of the future. The city's social contrasts remained, and the dreams of efficiency, harmony, and progress could only come true on a fantastic, temporary site—just like the American utopia of the Panama authors would remain limited to the Canal Zone.

While war and military organization had been used as metaphors for the new society, the evolving conflict in Europe added a sobering note of disillusionment to this vision. The exposition closed on December 4, 1915, and it was the last of its kind. In the United States, colossal world's fairs returned only in the 1930s—in a decade when, in conjunction with the Depression and another looming war, collectivist impulses and technological visions resurfaced in new shapes and forms. At the New York World's Fair in 1939, the large corporations assumed the role of the state in imagining the future. Another exposition was held simultaneously on San Francisco's Treasure Island, two years after the completion of the Golden Gate Bridge. Supported by George Creel, the former head of the Committee on Public Information during World War I, it unfolded its own vision of a renewed American West, turning once again toward the Pacific, but it attracted less attention than its East Coast rival.[127]

World's fairs are sites of contest, and the PPIE is a case in point. Due to their confusing symbolism and antiquated design, the social Darwinist sculptures and murals celebrating the achievement of the Panama Canal met frequent criticism. Live displays of foreign people drew little attention because they had nothing new to offer. Concessions equipped with modern features, however, proved successful. Most significantly, the Panama Canal exhibit not only attracted large crowds of visitors but also functioned as an interpretive device. Like the visual images from the Isthmus, the cleverly constructed miniature perpetuated the "engineered view" of a new American society based on the model of the Canal Zone.

Conclusion

Visiting a Construction Site

A few years before its centennial, the Panama Canal essentially remains the same structure that was completed in 1914. Even
though some ships had already been too big for the passage by the
1930s, few changes were made in the following decades. Only recently,
the former Culebra Cut was substantially widened to allow two-way
transits of large commercial vessels. It will take exactly a hundred years
to heave the Canal into a new era. By 2014 a new lane of traffic along the
existing seaway and a new set of locks are expected to open, doubling its
capacity. The groundbreaking ceremony for the expansion took place
on September 3, 2007—attended, once again, by former U.S. president
Jimmy Carter.

There is no simple answer to the question whether the waterway
ever earned a return on the investment it took to build it. It took many
lives, it forcefully transformed a large area of land and its people, and it
shaped the country of Panama, its culture and economy. Throughout
the twentieth century, the Canal served as a reminder of U.S. presence
and influence in the region. Then, on New Year's Eve of 1999, it was
handed over to Panama and placed under its control.

Roosevelt's justification of American involvement in the creation of
Panama, as laid out in his messages to Congress, framed the role of the
United States as the "policeman" of the world for decades to come. In

the present century, the ongoing conflict in Iraq underscores how exceptional the return of the Canal, settled in 1977, actually was. It is crucial
to remember, however, that this decision had been reached only after
one of the most heated political debates in American history—with the
effect of dividing the public and facilitating the rise of conservatism in
the 1980s. The debate also demonstrated that the Canal had always
remained an ambiguous terrain, a foreign yet familiar place deeply in-
grained in the American consciousness.

In this book, I focused on the Panama Canal as a construction site of
social visions, which so far has been a blind spot in the cultural history of
the American empire. Inevitably, the narrative at times resembled the
story of the physical building of the waterway. The Canal was a colossal
work, and like many other novelties at the beginning of the twentieth
century, it was baffling. It required explanations and interpretations, and
these were provided in the books and articles written by journalists,
travel writers, and officials involved in its making: the Panama authors.
Their audience was the American middle class. At the dawn of the modern age, as industrialization, consumption, immigration, and expansion
transformed Western societies, as gender roles and work patterns were
shifting, this audience was confronted with the task of envisioning the future, of escaping drift and achieving mastery. Driven by both anticipa-
tion and anxiety, it was engaged in a search for order. The building of
the Panama Canal became one of the manifestations of this search.

I regard the writings and images created by the Panama authors not
only as sources through which the historian gains access to the "reality"
of history: they shaped and determined this reality. They were constituents of a discourse relating the construction of the Panama Canal to the
challenges with which Americans were faced in their own country. For
example, I have analyzed President Roosevelt not mainly as a political
actor but as an "engineer" of the enterprise, as an interpreter influenc-
ing other interpretations, including my own. Not surprisingly, the com-
pilation of these statements about the Panama Canal revealed many
contradictions. The writers and policymakers imagined a perfect society
on the Isthmus, but they could reach perfection only through reduction
and exclusion, in the same way the utopian authors had conceived their
fictional worlds. They diminished the contribution of West Indian
workers and depicted the Canal Zone as a model state of white Americans. They celebrated a male society unhampered by the influence of
women. And they praised a collectivist state but took care to circumvent

the suspicious label "socialism." Their strategies reflect the contested nature of discourses. Synthesis and omission form the boundaries of what is said or imagined about a certain discursive subject. At the same time, by its very existence, this process acknowledges the possibility of resistance and the articulation of difference.

Many of the meanings attached to the building of the seaway resembled those evoked by earlier technological projects such as the Erie Canal and the Brooklyn Bridge or by ensuing endeavors such as those of the Tennessee Valley Authority. They all shared an appeal to civic unity, the belief in moral and technological progress, universal peace, and concerted action. The conception and rhetoric of the Panama-Pacific International Exposition, held to celebrate the opening of the Panama Canal, was similar to earlier world's fairs such as the Columbian Exposition in Chicago, even though twenty-two years had passed since then. And, of course, what seems original to American history finds its parallels in other parts of the world or, as recent transnational studies have shown, has been influenced by events and developments elsewhere. These correlations do not reduce the significance of the Canal project; on the contrary, they reinforce the messages expressed in its interpretation. Politicians, popular authors, and managers of the San Francisco World's Fair took advantage of the public's fascination with technological projects and their visual appeal and employed the building of the Canal as their instrument to restore the city of San Francisco, the American nation, and white civilization as a whole.

At its core, this limitless empire was not a territorial empire but a social, technological, and moral one. To Alfred Thayer Mahan, Brooks Adams, Frederick Jackson Turner, and Theodore Roosevelt, expansion abroad was not a goal in itself. The dominance over other people and races, markets, and strategic networks such as the oceans served to assure American society that it was capable of coping with the changes with which it was confronted. The phrase "bringing order out of chaos," which Roosevelt had applied to the former Spanish colonies now under U.S. control, was also a call for domestic action. Despite its reactionary reverberations, the expansionists' program was based on the decidedly modern, neo-Lamarckian belief that bodies, people, and nations could be remade in an evolutionary, transformative, and ultimately technological act. Thus, Roosevelt's visit to the Canal Zone symbolized the turnaround of the project from an ill-fated venture into a success story.

Although the Spanish-American War provided a moment of self-assurance (and equipped the United States with a large colony), the building of the Panama Canal was the more significant and yet overlooked project of forging a new national identity. It took place on a foreign territory but was presented as a logical extension of the United States. It served as an experiment for the future American society and addressed all aspects of this social body: technology, work, health, consumption, management, and administration—working together "in perfect operation." The panoramic view expressed in many of the Canal representations, perhaps most powerfully in the miniature at the San Francisco World's Fair, reflected this need for synthesis. The building of the Panama Canal was an integrated scheme, a master project, for a new America. Fitting the description of William James's "moral equivalent of war," it was more uplifting than physical battle and more authentic than a fictional utopia. The dichotomies embedded in this discourse—peace and war, harmony and aggression, national and international, past and future—were well suited for the interpretation of the Canal as a universal metaphor. At the same time, they reflected underlying uncertainties and the struggle for control.

The fact that the Panama Canal was a public endeavor became one of its most significant attributes. In contrast to the French, who represented the waste and immorality of an unchecked capitalism, the American management of the Canal Zone was seen as efficient, patriotic, manly, and authoritarian. Chief Engineer George W. Goethals became the emblem of a new type of social leader, the engineer-soldier, the benevolent despot. This view of the venture on the Isthmus corresponded to the efforts of reformers in the United States to centralize power in the hands of the state. The New Intellectuals shared the enthusiasm about the implications of the Canal project and remained divided on the need for democratic participation in a perfect society. The completion of the waterway coincided with the eruption of the war in Europe, and the impending American mobilization promised the chance of applying the lessons of Panama to the United States. Mastery seemed at hand. Instead, drift turned into disillusion. The same intellectuals, by now political advisers, compromised their radical ideas and eventually lost trust in the benevolence of the state. The war economy bolstered the ties between business and government. In the 1920s, the individualist ethos of private initiative gained new strength, but collectivist impulses within the American middle class survived, only to resurface in the New Deal and the rise of the corporation as a new social power.

In the twenty-first century, the age of globalization, the expectation that progress will eventually result in some kind of final, fixed order may seem naive, even though it enjoyed a brief moment of popularity immediately after the end of the cold war. The public sphere, competing with multinational corporations and nongovernmental organizations, is struggling to find its role in an emerging civil society. The murderous totalitarian regimes of the last century and the rise of a global capitalist system undoubtedly contributed to its descent. For the journalist Walter Lippmann, disillusionment came even earlier. The faith in the power of the state, as expressed in Edward Bellamy's novel and the interpretations of the Panama Canal, was rooted in a desire for the fulfillment of history and the notion that utopia was within reach. As long as this belief is based on the exclusion and discrimination of other people, it deserves to be shattered, then and now.

Notes

Approaching the Panama Canal: An Introduction

1. *New York Times*, Dec. 15, 1999, A1 and A14.

2. Amy Kaplan, "'Left Alone with America': The Absence of Empire in the Study of American Culture," in *Cultures of United States Imperialism*, ed. Amy Kaplan and Donald E. Pease (Durham, NC: Duke University Press, 1993), 11.

3. Robert W. Rydell, *All the World's a Fair: Visions of Empire at American International Expositions, 1890–1945* (New York: Hill and Wang, 1984).

4. Bill Brown, "Science Fiction, the World's Fair, and the Prosthetics of Empire, 1910–1915," in Kaplan and Pease, *Cultures of United States Imperialism*, 129–163. Brown focuses on the Panama-Pacific International Exposition in San Francisco held to celebrate the opening of the Canal. For a further discussion, see chap. 5.

5. This book is a revised version of the author's dissertation; see Alexander Missal, "'In Perfect Operation': American Social Visions and the Building of the Panama Canal, 1900–1915" (Ph.D. diss., University of Cologne, 2006). For an earlier outline, see Missal, "'In Perfect Operation': Social Vision and the Building of the Panama Canal," in *Dreams of Paradise, Visions of Apocalypse: Utopia and Dystopia in American Culture*, ed. Jaap Verheul (Amsterdam: VU University Press, 2004), 69–77.

6. George Frost Kennan, *American Diplomacy, 1900–1950* (Chicago: University of Chicago Press, 1951).

7. William Appleman Williams, *The Tragedy of American Diplomacy* (1959; New York: W. W. Norton, 1991). Williams revised his study twice, incorporating the lessons of the Cuba Crisis and the Vietnam War. On his influence, see Bradford Perkins, "The Tragedy of American Diplomacy Twenty-Five Years After," *Reviews in American History* 12 (March 1984): 1–18, and Lloyd C. Gardner, ed., *Redefining the Past: Essays in Diplomatic History in Honor of William Appleman Williams* (Corvallis: Oregon State University Press, 1986).

8. Walter LaFeber, *The New Empire: An Interpretation of American Expansion, 1860–1898* (Ithaca, NY: Cornell University Press, 1963). His expanded and updated study, *The Cambridge History of American Foreign Relations*, vol. 2: *The American Search for Opportunity, 1865–1913* (Cambridge, UK: Cambridge University Press, 1993), may be considered the canonized version of *The New Empire*.

9. Gilbert M. Joseph, "Toward a New Cultural History of U.S.-Latin American Relations," in *Close Encounters of Empire: Writing the Cultural History of U.S.-Latin American Relations*, ed. Gilbert M. Joseph et al. (Durham, NC: Duke University Press, 1998), 4.

10. An example of this trend is David Healy, *Drive to Hegemony: The United States in the Caribbean, 1898–1917* (Madison: University of Wisconsin Press, 1988). See also Thomas J. McCormick, "Drift or Mastery? A Corporatist Synthesis for American Diplomatic History," *Reviews in American History* 10 (Dec. 1982): 318–330.

11. See Gerstle Mack, *The Land Divided: A History of the Panama Canal and Other Isthmian Canal Projects* (New York: Alfred A. Knopf, 1944), and Miles P. DuVal Jr., *And the Mountains Will Move: The Story of the Building of the Panama Canal* (Stanford, CA: Stanford University Press, 1947). Mack was known as a biographer of artists, DuVal was a navy officer. Other works include David Howarth, *Panama: Four Hundred Years of Dreams and Cruelty* (New York: McGraw-Hill, 1966), and Donald Gordon Payne [Ian Cameron], *The Impossible Dream: The Building of the Panama Canal* (New York: William Morrow, 1972). The publication of David McCullough's *The Path between the Seas: The Creation of the Panama Canal, 1870–1914* (New York: Simon and Schuster, 1977) was timed perfectly in the midst of the "great debate" on the Canal, but the author notes that he had embarked on his research prior to the controversy (13).

12. Walter LaFeber, *The Panama Canal: The Crisis in Historical Perspective* (1978; Oxford: Oxford University Press, 1989). The second edition was published on the occasion of the seventy-fifth anniversary of the Canal's opening.

13. Matthew Parker, *Panama Fever: The Battle to Build the Canal* (London: Hutchinson, 2007).

14. John Major, *Prize Possession: The United States and the Panama Canal, 1903–1979* (Cambridge, UK: Cambridge University Press, 1993), like Parker's the work of a British historian, deserves the first mentioning. Other general analyses include Michael L. Conniff, *Panama and the United States: The Forced Alliance* (Athens: University of Georgia Press, 1992), and Almon R. Wright, *Panama: Tension's Child, 1502–1989* (New York: Vantage Press, 1990).

15. Neither the return of the Canal Zone in 1999/2000 nor the centennial of Panamanian "independence" in 2003 prompted a surge in the English-language literature on the subject. The National Museum of American History in Washington, D.C., organized the exhibition *Making the Dirt Fly! Building the Panama Canal: A Smithsonian Institution Libraries Exhibition, National*

Museum of American History, Nov. 20, 1999—Jan. 5, 2001, http://www.sil.si.edu/exhibitions/make-the-dirt-fly, accessed Sept. 22, 2007, on all aspects of the Canal construction.

16. Amy Kaplan, "Commentary: Domesticating Foreign Policy," *Diplomatic History* 18, no. 1 (1994): 99.

17. Akira Iriye, "Culture and Power: International Relations as Intercultural Relations," *Diplomatic History* 3 (Spring 1979): 115-128, is perhaps his earliest essay on the subject. In 1990, Iriye still speaks of "this third theme" with regard to the "cultural approach to diplomatic history." See his contribution "Culture" as part of "A Round Table: Explaining the History of American Foreign Relations," *Journal of American History* 77, no. 1 (1990): 99. Emily S. Rosenberg's *Spreading the American Dream: American Economic and Cultural Expansion, 1890-1945* (New York: Hill and Wang, 1982) built on the work of the New Left scholars.

18. For the characterization of this shift, see Nicholas B. Dirks et al., eds., *Culture/Power/History: A Reader in Contemporary Social Theory* (Princeton, NJ: Princeton University Press, 1994); Lynn Hunt, ed., *The New Cultural History* (Berkeley: University of California Press, 1989); Lynn Hunt, "Geschichte jenseits von Gesellschaftstheorie," in *Geschichte schreiben in der Postmoderne: Beiträge zur aktuellen Diskussion,* ed. Christoph Conrad and Martina Kessel (Stuttgart: Philipp Reclam jun., 1994), 98-122; and John E. Toews, "Intellectual History after the Linguistic Turn," *American Historical Review* 91 (Oct. 1987): 879-907. For the discussion in Germany, see Ute Daniel, *Kompendium Kulturgeschichte: Theorien, Praxis, Schlüsselwörter* (Frankfurt/Main: Suhrkamp, 2001); and Hans-Ulrich Wehler, *Die Herausforderung der Kulturgeschichte* (München: Beck, 1998), a halfhearted endorsement by the most prominent proponent of *Sozialgeschichte* (social history), which focuses on political, economic, and social institutions and structures.

19. Ricardo D. Salvatore, "The Enterprise of Knowledge: Representational Machines of Informal Empire," in Joseph et al., *Close Encounters of Empire,* 74. I would add that the flow goes from south to north (and many other directions) as well. The influence of the *cultural turn* on the practice of diplomatic history has been discussed in depth in "A Round Table," 93-180, and in a debate involving numerous contributors in *Diplomatic History* 18, no. 1 (1994): 59-124. It is reflected in the revised edition of Michael J. Hogan and Thomas G. Paterson, *Explaining the History of American Foreign Relations* (1991; Cambridge, UK: Cambridge University Press, 2004). Also noteworthy are Jessica C. E. Gienow-Hecht and Frank Schumacher, eds., *Culture and International History* (New York: Berghahn Books, 2003), and Ursula Lehmkuhl, "Diplomatiegeschichte als internationale Kulturgeschichte: Theoretische Ansätze und empirische Forschung zwischen Historischer Kulturwissenschaft und Soziologischem Institutionalismus," *Geschichte und Gesellschaft* 27, no. 3 (2001): 394-423. For a critique of a one-sided focus on discourse analysis, see Melvyn P. Leffler, "New

Approaches, Old Interpretations, and Prospective Reconfigurations," *Diplomatic History* 19, no. 2 (1995): 173–196.

20. For contributions to this research trend with regard to U.S. history, see David Thelen, "Making History and Making the United States," *Journal of American Studies* 32, no. 3 (1998): 373–397; Thomas Bender, ed., *Rethinking American History in a Global Age* (Berkeley: University of California Press, 2002); "Empires and Intimacies: Lessons from (Post)Colonial Studies: A Round Table," *Journal of American History* 88 (Dec. 2001), 829–898; and Manfred Berg and Philipp Gassert, eds., *Deutschland und die USA in der Internationalen Geschichte des 20. Jahrhunderts: Festschrift für Detlef Junker* (Stuttgart: Franz Steiner Verlag, 2004), 21–97.

21. An almost ironic acknowledgment of this fact is the title *After the Imperial Turn: Thinking with and through the Nation,* ed. Antoinette Burton (Durham, NC: Duke University Press, 2003). Burton has assembled contributions exemplifying almost every research trend, though mainly with regard to the British Empire. For the popularity of the empire terminology, see also the controversial Niall Ferguson, *Colossus: The Price of American Empire* (New York: Penguin Press, 2004), and Amy Kaplan's Presidential Address to the American Studies Association, "Violent Belongings and the Question of Empire Today," *American Quarterly* 56 (March 2004): 1–18.

22. See, for example, Frank Schumacher, "The American Way of Empire: National Tradition and Transatlantic Adaptation in America's Search for Imperial Identity, 1898–1910," *Bulletin of the German Historical Institute* 31 (Fall 2002): 35–50, preceding Schumacher's forthcoming study *The American Way of Empire: The United States and the Quest for Imperial Identity, 1880–1920.*

23. Michael Hardt and Antonio Negri, *Empire* (Cambridge, MA: Harvard University Press, 2000). For a concise evaluation of *Empire* and its context, see Norbert Finzsch, "Von Wallerstein zu Negri: Sind die USA das 'neue' Rom?" in *Der 11. September 2001: Fragen, Folgen, Hintergründe,* ed. Sabine Sielke (Frankfurt/Main/Berlin: Lang, 2002), 159–171. For a conventional Marxist critique, see Atilio A. Boron, *"Empire" and Imperialism: A Critical Reading of Michael Hardt and Antonio Negri,* trans. Jessica Casiro (London: Zed Books, 2005).

24. Hardt and Negri, *Empire,* 167.

25. Ibid., 174.

26. See pp. 29–30. The title of Brooks Adams's book written in 1902, *The New Empire,* not only served as the inspiration for the naming of Walter LaFeber's above-cited work on American expansionism but also reverberates in Hardt's and Negri's description of "a new Empire with open, expanding frontiers, where power would be effectively distributed in networks," xiv. See Brooks Adams, *The New Empire: With an Appendix Containing a Chronological Survey from 4000 B.C. up till 1900* (1902; New York: Bergman, 1969).

27. See Stephen Frenkel, "Jungle Stories: American Representations of Tropical Panama," *Geographical Review* 86 (July 1996): 317–333, and his "Geographical Representations of the 'Other': The Landscape of the Panama Canal Zone," *Journal of Geographical History* 28, no. 1 (2002): 85–99, as well as the often speculative John Lindsay-Poland, *Emperors in the Jungle: The Hidden History of the U.S. in Panama* (Durham, NC: Duke University Press, 2003). Both authors stress the influence of racism on the social and environmental development in Panama and the Canal Zone.

28. See Velma Newton, *The Silver Men: West Indian Labor Migration to Panama, 1850–1914* (Mona, Kingston 7, Jamaica: Institute of Social and Economic Research, University of the West Indies, 1984); Michael L. Conniff, *Black Labor on a White Canal: Panama, 1904–1981* (Pittsburgh: University of Pittsburgh Press, 1985), and Julie Greene, "Spaniards on the Silver Roll: Labor Troubles and Liminality in the Panama Canal Zone, 1904–1914," *International Labor and Working-Class History* 66 (Fall 2004): 78–98.

29. See Michael Adas, *Dominance by Design: Technological Imperatives and America's Civilizing Mission* (Cambridge, MA: Belknap Press of Harvard University Press, 2006), esp. 185–198. On the cultural impact of the Canal on Latin American countries, see Ricardo D. Salvatore, "Imperial Mechanics: South America's Hemispheric Integration in the Machine Age," *American Quarterly* 58 (Sept. 2006): 662–691.

30. Alfred Charles Richard Jr., *The Panama Canal in American National Consciousness, 1870–1990* (New York: Garland, 1990), 213.

31. J. Michael Hogan, *The Panama Canal in American Politics: Domestic Advocacy and the Evolution of Policy* (Carbondale: Southern Illinois University Press, 1986). The author is not to be confused with Michael J. Hogan, a prominent scholar of American foreign relations.

32. Ibid., 46, and passim.

33. Richard, *Panama Canal in American National Consciousness*, 213, reports that at the time more than twenty-five book-length popular histories were published on the Canal building. For this study, I have focused on the following works: Willis John Abbot, *Panama and the Canal in Picture and Prose: A Complete Story of Panama, as Well as the History, Purpose and Promise of Its World-Famous Canal—the Most Gigantic Engineering Undertaking since the Dawn of Time* (New York: Syndicate, 1913); Charles Francis Adams, *The Panama Canal Zone: An Epochal Event in Sanitation* (Boston: Massachusetts Historical Society, 1911); Ralph Emmett Avery, *America's Triumph at Panama: Panorama and Story of the Construction and Operation of the World's Giant Waterway from Ocean to Ocean* (Chicago: Regan Printing House, 1913); Ralph Emmett Avery, *The Panama Canal and the Golden Gate Exposition: Authentic and Complete Story of the Building and Operation of the Great Waterway—the Eighth Wonder of the World* (New York: Leslie-Judge, 1915); Ira E. Bennett, *History*

of the Panama Canal: Its Construction and Builders (Washington, DC: Historical, 1915); Farnham Bishop, *Panama Past and Present* (New York: Century, 1913); Joseph Bucklin Bishop, *The Panama Gateway* (New York: Charles Scribner's Sons, 1913); Arthur Bullard [Albert Edwards], *Panama: The Canal, the Country and the People*, rev. ed. with additional chapters (1911; New York: Macmillan, 1914); Mary A. Chatfield, *Light on Dark Places at Panama: By an Isthmian Stenographer* (New York: Broadway, 1908); John O. Collins, *The Panama Guide* (Panama: Vibert and Dixon, 1912); Charles Harcourt Forbes-Lindsay, *The Isthmus and the Canal*, rev. ed. (1906; Philadelphia: John C. Winston, 1912); Harry A. Franck, *Zone Policeman 88: A Close Range Study of the Panama Canal and Its Workers* (New York: Century, 1913); John Foster Fraser, *Panama and What It Means* (London: Cassell, 1913); Frank A. Gause and Charles Carl Carr, *The Story of Panama: The New Route to India* (Boston: Silver, Burdett, 1912); Alfred B. Hall and Clarence L. Chester, *Panama and the Canal* (New York: Newson, 1910); Frederic Jennings Haskin, *The Panama Canal* (Garden City, NY: Doubleday, Page, 1913); Willis Fletcher Johnson, *Four Centuries of the Panama Canal* (New York: Cassell, 1907); *Joseph Pennell's Pictures of the Panama Canal: Reproductions of a Series of Lithographs Made by Him on the Isthmus of Panama, January–March, 1912, Together with Impressions and Notes by the Artist* (Philadelphia: J. B. Lippincott, 1912); Logan Marshall, *The Story of the Panama Canal* (Philadelphia: John C. Winston, 1913); Mary L. McCarty, *Glimpses of Panama and the Canal* (Kansas City: Tiernan-Dart Printing, 1913); William Lewis Nida, *Story of Panama and the Canal* (Chicago: Hall and McCreary, 1913); Walter Leon Pepperman, *Who Built the Panama Canal?* (London: J. M. Dent and Sons, 1915); William Rufus Scott, *The Americans in Panama* (New York: Statler, 1912); Hugh C. Weir, *The Conquest of the Isthmus: The Men Who Are Building the Panama Canal—Their Daily Lives, Perils, and Adventures* (New York: G. P. Putnam's Sons, 1909); Hugh C. Weir, *With the Flag at Panama: A Story of the Building of the Panama Canal* (Boston: W. A. Wilde, 1911); and A. W. Wyndham, *The Panama Canal* (New York: Howard F. Curtis, 1907). It should be noted that only three of the authors were female. This list does not include works that devote only one or two chapters to the Panama Canal nor the books and articles published by top Canal officials such as Chief Engineer George W. Goethals and Chief Sanitary Officer William C. Gorgas. For their full citations, refer to the notes in the respective chapters. For books and booklets published on the Panama-Pacific International Exposition in San Francisco 1915, see p. 252, n. 11. Based on the index of the *Readers' Guide to Periodical Literature*, vol. 1–5 (Minneapolis: H. W. Wilson, 1905–1919), I have also evaluated hundreds of articles on the Canal construction and the exposition in San Francisco—many by the same authors. A total figure was not recorded, but Healy, *Drive to Hegemony*, 92, mentions that the guide lists 546 entries on the Panama Canal between 1900 and 1914. I occasionally cite the *New York Times* but have generally refrained from examining newspaper reporting on the Panama Canal since most papers—in contrast to the national magazines—had local audiences and expressed partisan views, often based on the general political agenda of the publishing house.

34. J. M. Hogan, *Panama Canal*, 53–54.

35. Herbert Knapp and Mary Knapp, *Red, White, and Blue Paradise: The American Canal Zone in Panama* (San Diego: Harcourt Brace Jovanovich, 1984). The Knapps lived in the Canal Zone from 1963 to 1982. They present both their personal recollections as well as a historical sketch of the Canal building, based largely on the Panama authors. For a discussion, see p. 135.

36. B. Brown, "Science Fiction," 158, n19.

37. Amy Kaplan, *The Anarchy of Empire in the Making of U.S. Culture* (Cambridge, MA: Harvard University Press, 2002), 15.

38. Gail Bederman, *Manliness and Civilization: A Cultural History of Gender and Race in the United States, 1880–1917* (Chicago: University of Chicago Press, 1995), 13–14.

39. Walter Lippmann, *Drift and Mastery: An Attempt to Diagnose the Current Unrest* (1914; Madison: University of Wisconsin Press, 1985), 111–112.

40. James J. Connolly, "GAPE Bibliography: Progressivism," Jan. 31, 1997, available through http://www.h-net.org/~shgape, accessed Sept. 22, 2007. The literature on Progressivism and reform is abundant and keeps growing. On the controversial concept of a "movement," see Peter G. Filene, "An Obituary for the Progressive Movement," *American Quarterly* 22 (Spring 1970): 20–23, and Daniel Rodgers, "In Search of Progressivism," *Reviews in American History* 10 (Dec. 1982): 113–132. The most important contributions to the historiography include Richard Hofstadter, *Age of Reform: From Bryan to F.D.R.* (New York: Alfred A. Knopf, 1955); Gabriel Kolko, *The Triumph of Conservatism: A Reinterpretation of American History, 1900–1916* (Chicago: Quadrangle Books, 1963); Robert Wiebe, *The Search for Order, 1877–1920* (New York: Hill and Wang, 1967); James Weinstein, *The Corporate Ideal in the Liberal State, 1900–1918* (Boston: Beacon Press, 1968); Martin Sklar, *The Corporate Reconstruction of American Capitalism, 1890–1916: The Market, the Law, the Politics* (Cambridge, UK: Cambridge University Press, 1988); and Morton Keller, *Regulating a New Society: Public Policy and Social Change in America, 1900–1933* (Cambridge, MA: Harvard University Press, 1994). For a focus on gender and race, see Bederman, *Manliness and Civilization*. John Whiteclay Chambers II, *The Tyranny of Change: America in the Progressive Era, 1900–1917* (New York: St. Martin's Press, 1980), and more recently Michael McGerr, *A Fierce Discontent: The Rise and Fall of the Progressive Movement in America, 1870–1920* (New York: Free Press, 2003), and J. Michael Hogan, ed., *Rhetoric and Reform in the Progressive Era* (East Lansing: Michigan State University Press, 2003), are useful overviews. For comparative approaches that added a new perspective to the field, see Daniel Rodgers, *Atlantic Crossings: Social Politics in a Progressive Age* (Cambridge, MA: Belknap Press of Harvard University Press, 1998), and Axel R. Schäfer, *American Progressives and German Social Reform, 1875–1920:*

Social Ethics, Moral Control, and the Regulatory State in a Transatlantic Context (Stuttgart: Steiner, 2000).

41. Warren I. Susman, *Culture as History: The Transformation of American Society in the Twentieth Century* (New York: Pantheon Books, 1984), xxi.

42. On the role of national magazines, see Richard Ohmann, *Selling Culture: Magazines, Markets, and Class at the Turn of the Century* (London: Verso, 1996); Matthew Schneirow, *The Dream of a New Social Order: Popular Magazines in America, 1893–1914* (New York: Columbia University Press, 1994); and the classic Frank Luther Mott, *A History of American Magazines, 1885–1905* (Cambridge, MA: Belknap Press of Harvard University Press, 1957).

43. See Schneirow, *Dream of a New Social Order*, 5; Mott, *History of American Magazines*, 11; and Ohmann, *Selling Culture*, 223, 360.

44. Salvatore, "Enterprise of Knowledge," 82.

45. See *The National Cyclopedia of American Biography*, vol. 32 (New York: James T. White, 1945), 266.

46. Richard, *Panama Canal in American National Consciousness*, 214–215, fails to note that the authors were identical.

47. Charles Harcourt Forbes-Lindsay, *Bridge, and How to Play it* (Philadelphia: Penn, 1908), and William Rufus Scott, *The Itching Palm: A Study of the Habit of Tipping in America* (Philadelphia: Penn, 1916). Weir was also a novelist and movie writer who later collaborated with Irvin Thalberg. See his obituary in the *New York Times*, March 18, 1934, 35. Obituaries and articles from the *National Cyclopedia of American Biography* are often the only sources of information available on these authors. On Bullard, see pp. 152–153.

48. George Cotkin, *Reluctant Modernism: American Thought and Culture, 1880–1900* (New York: Twayne, 1992), 103.

49. See Abbot, *Panama and the Canal*, n.p.

50. Salvatore, "Enterprise of Knowledge," 74; Salvatore applies this term to the social construction of Latin America in general.

51. See Joseph Bucklin Bishop, *Theodore Roosevelt and His Time: Shown in His Own Letters* (New York: Charles Scribner's Sons, 1920).

52. McCullough, *Path between the Seas*, 536.

53. Memorandum by commission chairman Theodore P. Shonts, Sept. 7, 1905; Joseph Bucklin Bishop file; American Citizen Official Personal Folders, 1903–20; Records of the Panama Canal, Record Group (RG) 185, National Archives at College Park (NACP), Maryland.

54. Memorandum by Commission Secretary Joseph Bucklin Bishop to Shonts, n.d., Joseph Bucklin Bishop file; American Citizen Official Personal Folders, 1903–20, RG 185, NACP.

55. Clifford Geertz, "The Politics of Meaning," in *The Interpretation of Cultures: Selected Essays by Clifford Geertz* (New York: Basic Books, 1973), 312.

56. Clifford Geertz, "Thick Description: Toward an Interpretive Theory of Culture," in Geertz, *Interpretation of Cultures*, 5. Geertz acknowledges that he

bases parts of his concept and terminology on Max Weber. On the critique of Geertz's work and its application in the 1980s and 1990s, see Sherry B. Ortner, ed., *The Fate of "Culture": Geertz and Beyond* (Berkeley: University of California Press, 1999). For a summary of the evolving relationship between culture and history, see Christoph Konrad and Martina Kessel, "Blickwechsel: Moderne, Kultur, Geschichte," in *Kultur und Geschichte: Neue Einblicke in eine alte Beziehung* (Stuttgart: Philipp Reclam jun., 1998), 9–40.

57. Benedict Anderson, *Imagined Communities: Reflections on the Origin and Spread of Nationalism* (London: Verso, 1983), esp. 13, 15, 19.

58. Geertz, "Thick Description," 5.

59. Ibid., 9.

60. Alun Munslow, *Deconstructing History* (London: Routledge, 1997), 121.

61. See Robert F. Berkhofer Jr., *Beyond the Great Story: History as Text and Discourse* (Cambridge, MA: Belknap Press of Harvard University Press, 1995), 63.

62. Geertz, "Thick Description," 15.

63. Alun Munslow, *Discourse and Culture: The Creation of America, 1870–1920* (London: Routledge, 1992), 166.

64. See James Clifford, *The Predicament of Culture: Twentieth-Century Ethnography, Literature, and Art* (Cambridge, MA: Harvard University Press, 1988), 264.

65. See Berkhofer, *Beyond the Great Story*, 80–81.

66. On the notion of the discursive subject being no longer "a free subject of thought or action," see Edward W. Said, *Orientalism* (New York: Vintage Books, 1979), 3. This work is one of the earliest attempts (and probably the best known) to apply Foucault's concepts in a terrain untouched by the French philosopher's own historical studies.

67. See Michel Foucault, *The History of Sexuality: An Introduction* [Histoire de la sexualité, I: La volonté de savoir], trans. Robert Hurley (1976; London: Penguin, 1978), 12; Foucault stresses that power is only partially (i.e., locally and tactically) exerted through repression.

68. Munslow, *Discourse and Culture*, 164.

69. See Bederman, *Manliness and Civilization*, 24, as well as Foucault, *History of Sexuality*, 100–102, and *The Archaeology of Knowledge* [Archéologie du savoir], trans. A. M. Sheridan Smith (London: Routledge, 1972), 151, where he states that "such a contradiction, far from being an appearance or accident of discourse, far from being that from which it must be freed if its truth is at last to be revealed, constitutes the very law of its existence: it is on the basis of such a contradiction that discourse emerges, and it is in order both to translate it and to overcome it that discourse begins to speak."

70. See Michel Foucault, *The Order of Things: An Archaeology of the Human Sciences* [Mots et les choses] (London: Tavistock, 1970), *L'ordre du discourse: Leçon inaugurale au Collège de France prononcée le 2 décembre 1970* (Paris: Gallimard, 1971), and *The Archaeology of Knowledge*, esp. 128–131, 139–140. Clifford, *Predicament of Culture*, 268–269, for instance, criticizes Said's reliance on individual authors in

his application of Foucault's scheme. For a discussion of Foucault's influence on historians, see Colin Jones and Roy Porter, eds., *Reassessing Foucault: Power, Medicine, and the Body* (London: Routledge, 1994), and Jürgen Martschukat, ed., *Geschichte schreiben mit Foucault* (Frankfurt/Main: Campus, 2002).

71. Norbert Finzsch, "Conditions of Intolerance: Racism and the Construction of Social Reality," *Historical Social Research* 22, no. 1 (1997): 25.

Chapter 1. Logistics of Expansion

1. For the following paragraphs, see also Michael L. Conniff, *Panama and the United States: The Forced Alliance* (Athens: University of Georgia Press, 1992), 7–67; John Major, *Prize Possession: The United States and the Panama Canal, 1903–1979* (Cambridge, UK: Cambridge University Press, 1993), 9–33; Walter LaFeber, *The Panama Canal: The Crisis in Historical Perspective* (1978; Oxford: Oxford University Press, 1989), 3–22, and David McCullough, *The Path between the Seas: The Creation of the Panama Canal, 1870–1914* (New York: Simon and Schuster, 1977), 19–341.

2. Alexander de Humboldt, *Political Essay on the Kingdom of New Spain: Translated from the Original French by John Black*, vol. 1 (London: Longman, Hurst, Rees, Orme and Brown, 1811), 18.

3. Conniff, *Panama and the United States*, 51.

4. Ibid., 33.

5. William Crawford Gorgas, *Sanitation in Panama* (New York: D. Appleton, 1915), 149.

6. See Matthew Parker, *Panama Fever: The Battle to Build the Canal* (London: Hutchinson, 2007), 160–163.

7. Matt K. Matsuda, *Empire of Love: Histories of France and the Pacific* (Oxford: Oxford University Press, 2005), 57. The author discusses the interpretations of a future canal from a French perspective. See also pp. 68 and 78.

8. For a history of the Compagnie Nouvelle du Canal de Panama, see James M. Skinner, *France and Panama: The Unknown Years, 1894–1908* (New York: P. Lang, 1989).

9. See *Scientific American*, Jan. 18, 1902, 34.

10. For the following paragraphs, see also Walter LaFeber, *The New Empire: An Interpretation of American Expansion, 1860–1898* (Ithaca, NY: Cornell University Press, 1963), 13–24.

11. Robert Wiebe, *The Search for Order, 1877–1920* (New York: Hill and Wang, 1967), 235.

12. J. Michael Hogan, *The Panama Canal in American Politics: Domestic Advocacy and the Evolution of Policy* (Carbondale: Southern Illinois University Press, 1986), 19. In an isolated chapter of *The New Empire*, LaFeber discusses the ideas of the expansionists (62–101), but neglects to re-evaluate their complex agenda in his book on the Panama Canal.

13. LaFeber, *New Empire*, 58.

14. Alfred Thayer Mahan, *The Influence of Sea Power upon History, 1660–1783* (1890; New York: Dover, 1987).

15. *Atlantic Monthly* 6 (Oct. 1890), 563. Roosevelt wrote the article anonymously.

16. Alfred Thayer Mahan, *The Interest of America in Sea Power: Present and Future* (1897; Port Washington, NY: Kennikat Press, 1970), 25, 124. The book contains a collection of essays published between 1890 and 1897 in prominent magazines.

17. Alfred Thayer Mahan, "The Isthmus and Sea Power," *Atlantic Monthly* 72 (Oct. 1893), 459, 460.

18. Mahan, *Interest of America in Sea Power*, 22.

19. See Farnham Bishop, *Panama Past and Present* (New York: Century, 1913); Arthur Bullard [Albert Edwards], *Panama: The Canal, the Country and the People*, rev. ed. with additional chapters (1911; New York: Macmillan, 1914), 463; Frank A. Gause and Charles Carl Carr, *The Story of Panama: The New Route to India* (Boston: Silver, Burdett, 1912), 276; Frederic Jennings Haskin, *The Panama Canal* (Garden City, NY: Doubleday, Page, 1913), 10, and Logan Marshall, *The Story of the Panama Canal* (Philadelphia: John C. Winston, 1913), 201. A contemporary school book explained: "It took so much time to do this that our government decided, whatever the cost, it must build a canal across the Isthmus as a future safeguard to our naval defences as well as for the great commercial advantages which would follow." William Lewis Nida, *Story of Panama and the Canal* (Chicago: Hall and McCreary, 1913), 14.

20. Mahan, *Interest of America in Sea Power*, 260–261. On the ambivalence of Mahan's quest, see also Anders Stephanson, *Manifest Destiny: American Expansion and the Empire of Right* (New York: Hill and Wang, 1995), 86.

21. Ibid., 95–96.

22. Brooks Adams, *The Law of Civilization and Decay: An Essay on History* (1896; New York/London: Macmillan, 1898), xi.

23. Ibid., 351.

24. Ibid., xi.

25. Brooks Adams, *New Empire*, 208. See also LaFeber, *New Empire*, 84, and T. J. Jackson Lears, *No Place of Grace: Antimodernism and the Transformation of American Culture, 1880–1920* (New York: Pantheon Books, 1981), 132–136.

26. On social Darwinism, see the classic but controversial Richard Hofstadter, *Social Darwinism in American Thought* (1944; Boston: Beacon Press, 1955), as well as Robert C. Bannister, *Social Darwinism: Science and Myth in Anglo-American Social Thought* (Philadelphia: Temple University Press, 1979), and Carl N. Degler, *In Search of Human Nature: The Decline and Revival of Darwinism in American Social Thought* (Oxford: Oxford University Press, 1991).

27. Gail Bederman, *Manliness and Civilization: A Cultural History of Gender and Race in the United States, 1880–1917* (Chicago: University of Chicago Press, 1995), 25.

28. Degler, *In Search of Human Nature*, 14. On the interdependence of racism at home and abroad, see Herbert Shapiro, "Racism and Empire: A Perspective on a New Era of American History," in *Identity and Intolerance: Nationalism, Racism, and Xenophobia in Germany and the United States*, ed. Norbert Finzsch and Dietmar Schirmer (Cambridge, UK: Cambridge University Press, 1998), 166–168.

29. On Strong and Fiske, see LaFeber, *New Empire*, 72–80, 99–100, and for a revisionist evaluation, Bannister, *Social Darwinism*, 228–229.

30. Mahan, *Interest of America in Sea Power*, 55. In Mahan's and Roosevelt's eyes, the only other peoples deserving respect for their evolutionary achievements were the Asian nations, especially Japan. They were still hesitant, though, to encourage a mixing of the "races," for instance, through Asian immigration to the West Coast. "In the present state of world's progress it is highly inadvisable that peoples in wholly different stages of civilization, or of wholly different types of civilization even although both equally high, shall be thrown into intimate contact," Roosevelt later wrote in his memoirs. Theodore Roosevelt, *An Autobiography* (1913; New York: Da Capo Press, 1985), 393. On Anglo-Saxonism, see also Stuart Anderson, *Race and Rapprochement: Anglo-Saxonism and Anglo-American Relations, 1895–1904* (East Brunswick, NJ: Associated University Press, 1981).

31. See Degler, *In Search of Human Nature*, 20–21.

32. Mahan, *Interest of America in Sea Power*, 125.

33. On the concept of the frontier, see Richard Slotkin, *Gunfighter Nation: The Myth of the Frontier in Twentieth-Century America* (New York: Atheneum, 1992), and David M. Wrobel, *The End of American Exceptionalism: Frontier Anxiety from the Old West to the New Deal* (Lawrence: University Press of Kansas, 1993).

34. Roosevelt to Turner, Feb. 10, 1894, in *The Letters of Theodore Roosevelt*, ed. Elting E. Morison, vol. 1 (Cambridge, MA: Harvard University Press, 1951), 363.

35. Frederick Jackson Turner, "The Problem of the West," *Atlantic Monthly* 78 (Sept. 1896), 289.

36. Ibid., 296.

37. Wiebe, *Search for Order*, 189.

38. On explanations of TR's long-lasting legacy, see David Greenberg, "Bhagwan Teddy: Explaining the Cult of Theodore Roosevelt," *Slate*, March 28, 2002, www.slate.com/id/2063795/, accessed Sept. 22, 2007, who states that both Presidents Bush as well as Bill Clinton invoked the memory of their predecessor. Edmund Morris has so far published *The Rise of Theodore Roosevelt* (New York: Coward, McCann and Geoghegan, 1979) and *Theodore Rex* (New York: Random House, 2001), the first two parts of a trilogy intended for a general audience.

39. Herbert George Wells, *The Future in America: A Search for Realities* (New York: Harper and Brothers, 1906), 253.

40. Theodore Roosevelt, *The Winning of the West*, 4 vols. (1889–1896; Lincoln: University of Nebraska Press, 1995).

41. Roosevelt to Mahan, May 3, 1897, in *Letters of Theodore Roosevelt*, 1:607.

On the friendship of the two historians and strategists, see Richard W. Turk, *The Ambiguous Relationship: Theodore Roosevelt and Alfred Thayer Mahan* (Westport, CT: Greenwood Press, 1987).

42. Roosevelt to Lodge, Sept. 23, 1901, in *Letters of Theodore Roosevelt*, 3:150.

43. On TR's personality, with special regard to his gender and his racial and foreign politics, see the chapter "Theodore Roosevelt: Manhood, Nation, and 'Civilization'" in Bederman, *Manliness and Civilization*, 170–215; Frank Ninkovich, "Theodore Roosevelt: Civilization as Ideology," *Diplomatic History* 10 (Summer 1986): 221–245; Arnaldo Testi, "The Gender of Reform Politics: Theodore Roosevelt and the Culture of Masculinity," *Journal of American History* 81 (March 1995): 1509–1533; Gary Gerstle, "Theodore Roosevelt and the Divided Character of American Nationalism," *Journal of American History* 86 (Dec. 1999): 1280–1307; and Emily S. Rosenberg, *Financial Missionaries to the World: The Politics and Culture of Dollar Diplomacy, 1900–1930* (Cambridge, MA: Harvard University Press, 1999), 35–41. Richard H. Collin, in his works *Theodore Roosevelt, Culture, Diplomacy, and Expansion: A New View of American Imperialism* (Baton Rouge: Louisiana State University Press, 1985) and *Theodore Roosevelt's Caribbean: The Panama Canal, the Monroe Doctrine, and the Latin American Context* (Baton Rouge: Louisiana State University Press, 1990), unfolds a revisionist argument in defense of TR's policies.

44. See John F. Kasson, *Houdini, Tarzan, and the Perfect Man: The White Male Body and the Challenge of Modernity in America* (New York: Hill and Wang, 2001).

45. See Joy F. Kasson, *Buffalo Bill's Wild West: Celebrity, Memory, and Popular History* (New York: Hill and Wang, 2000).

46. For sociocultural conceptions of the male gender, see R. W. Connell, *Masculinities* (Berkeley: University of California Press, 1995). For comprehensive historical analyses, see E. Anthony Rotundo, *American Manhood: Transformations in Masculinity from the Revolution to the Modern Era* (New York: Basic Books, 1993), and Michael Kimmel, *Manhood in America: A Cultural History* (New York: Free Press, 1996). As Bederman, *Manliness and Civilization*, 11, notes, the terming of a "crisis" of masculinity in the 1890s runs the risk of conveying the impression that there exists a kind of universal, positive definition of manhood that can (and should) be restored.

47. James Gilbert, *Men in the Middle: Searching for Masculinity in the 1950s* (Chicago: University of Chicago Press, 2005), 23–24. Gilbert's study includes a historiography of the most important contributions to the field (15–33).

48. John F. Kasson, *Houdini*, 75.

49. On Wister's novel, see Kim Townsend, *Manhood at Harvard: William James and Others* (New York: W. W. Norton, 1996), 268–273.

50. Kaplan, *Anarchy of Empire*, 99.

51. In the public discourse, this idea of self-invented manhood was increasingly referred to as "masculinity." See Bederman, *Manliness and Civilization*, 18,

and also Michael J. Kimmel, "Consuming Manhood: The Feminization of American Culture and the Recreation of the Male Body, 1832–1920," in *The Male Body: Features, Destinies, Exposures*, ed. Laurence Goldstein (Ann Arbor: University of Michigan Press, 1994), 12–42.

52. Theodore Roosevelt, "The Strenuous Life," in *The Strenuous Life: Essays and Addresses* (New York: Century, 1902), 4.

53. See George Cotkin, *Reluctant Modernism: American Thought and Culture, 1880–1900* (New York: Twayne, 1992), 135–136; Cotkin notes that Veblen assigned these nonproductive activities to a regressive stage in the course of evolution, while he admired the workmanship of skilled laborers and engineers. Thorstein Veblen, *The Theory of the Leisure Class* (1899; New Brunswick, NJ: Transaction, 1992).

54. Roosevelt, "Strenuous Life," 6.

55. Roosevelt, "Character and Success," in *Strenuous Life*, 121. On Roosevelt's praise for the book, see Roosevelt to Wheeler, Feb. 2, 1900, in *Letters of Theodore Roosevelt*, 2:1205. On Mahan's historic references, see *Interest of America in Sea Power*, 247, 254.

56. See Bederman, *Manliness and Civilization*, 181.

57. The amount of literature on the Spanish-American War is impressive. The classic work is Ernest R. May, *Imperial Democracy: The Emergence of America as a Great Power* (New York: Harcourt, Brace and World, 1961). For a gender perspective, see Kristin L. Hoganson, *Fighting for American Manhood: How Gender Politics Provoked the Spanish-American and Philippine-American Wars* (New Haven, CT: Yale University Press, 1998). An overview of the historiography is provided by Anne Cipriano Venzon and Martin Gordon, eds., *America's War with Spain: A Selected Bibliography* (Lanham, MD: Scarecrow Press, 2003).

58. See Gerstle, "Theodore Roosevelt," 1287. On Roosevelt's involvement in the war, see ibid., 1286–1295, and Amy Kaplan, "Black and Blue on San Juan Hill," in *Cultures of United States Imperialism*, ed. Kaplan and Pease, 219–236.

59. See Gerstle, "Theodore Roosevelt," 1292. As a consequence, African Americans failed to gain access to officer status and were assigned only minor combat roles in World War I.

60. Robert Dallek, "National Mood and Foreign Policy: A Suggestive Essay," *American Quarterly* 34 (Dec. 1982), 341.

61. Roosevelt, "Strenuous Life," 12.

62. See ibid., 14: "Our army needs complete reorganization,—not merely enlarging,—and the reorganization can only come as the result of legislation."

63. See David Axeen, "'Heroes of the Engine Room': American 'Civilization' and the War with Spain," *American Quarterly* 36 (Fall 1984), 484. According to Axeen, 498, Roosevelt had analyzed the Spanish-American War along the same lines, stating that "the Spanish too had courage. What they lacked were energy, training, forethought."

64. Roosevelt, "Strenuous Life," 6.

65. Ibid., 7–8.

66. Ibid., 9. Roosevelt's wording bears a striking resemblance to the subtitle of Wheeler's biography of Alexander the Great, *The Merging of the East and West*.

67. Rosenberg, *Financial Missionaries*, 32, 60–61.

68. Hoganson, *Fighting for American Manhood*, 181–187. On the events in the Philippines, see p. 146 for a more detailed discussion.

69. Ibid., 199. In contrast to Hoganson's emphasis on gender politics, Eric T. Love, in *Race over Empire: Racism and U.S. Imperialism, 1865–1900* (Chapel Hill: University of North Carolina Press, 2004), argues that the policymakers' ambivalent views on racial issues played a large part in their decision to renounce annexation.

70. For the following paragraphs, see also LaFeber, *Panama Canal*, 22–36; Major, *Prize Possession*, 34–63; and Michael LaRosa and Germán R. Mejía, eds., *The United States Discovers Panama: The Writings of Soldiers, Scholars, Scientists, and Scoundrels, 1850–1905* (Lanhan, MD: Rowman and Littlefield, 2004), 268.

71. See "Message Communicated to the Two Houses of Congress at the Beginning of the Second Session of the 58th Congress," Dec. 7, 1903, in *Addresses and Presidential Messages of Theodore Roosevelt, 1902–1904: With an Introduction by Henry Cabot Lodge* (New York: G. P. Putnam's Sons, 1904), 421, and Roosevelt, *Autobiography*, 529–530. The Panama authors repeated his statement. See F. Bishop, *Panama Past and Present*, 133, and Ira E. Bennett, *History of the Panama Canal: Its Construction and Builders* (Washington, DC: Historical Publishing Company, 1915), 228.

72. F. Bishop, *Panama Past and Present*, 147.

73. Roosevelt to Kermit Roosevelt, Nov. 4, 1903, in *Letters of Theodore Roosevelt*, 3:644.

74. Article 3, reprint of the Treaty in LaFeber, *Panama Canal*, 226.

75. Ibid., 30.

76. Article 7, reprint of the Treaty in LaFeber, *Panama Canal*, 226.

77. Ibid., 36.

78. Philippe Bunau-Varilla, *Panama: The Creation, Destruction, and Resurrection* (London: Constable, 1913), 378. Ovidio Diaz Espino, in his book *How Wall Street Created a Nation: J. P. Morgan, Teddy Roosevelt, and the Panama Canal* (New York: Four Walls Eight Windows, 2001), argues that Bunau-Varilla was part of a Wall Street syndicate that had bought the shares of the new French Canal company cheaply and then sold them to the government for the agreed $40 million. This assertion, first made by a newspaper, had caused a lawsuit in 1908. See McCullough, *Path between the Seas*, 385.

79. See Conniff, *Panama and the United States*, 79.

80. See LaFeber, *Panama Canal*, 38–39.

81. "Message Communicated to the Two Houses of Congress," Jan. 4, 1904, in *Addresses and Presidential Messages*, 458.

82. See LaFeber, *Panama Canal*, 42.

83. "Message Communicated to the Two Houses of Congress," Jan. 4, 1904, in *Addresses and Presidential Messages*, 452.

84. Joseph Bucklin Bishop, *Theodore Roosevelt and His Time: Shown in His Own Letters*, vol. 1 (New York: Charles Scribner's Sons, 1920), 308. For the different versions, see James F. Vivian, "The 'Taking' of the Panama Canal Zone: Myth and Reality," *Diplomatic History* 4 (Winter 1980): 95–98. Bishop's rendition is the most prominent one, but it is unclear whether Roosevelt said "Canal," "Isthmus," or "Panama Canal" instead of "Canal Zone" or whether he meant to say "I took a trip to the Canal," referring to his 1906 visit to the Isthmus discussed later in this chapter.

85. See J. M. Hogan, *Panama Canal*, 62. TR justified his actions in a magazine article appearing a few months after his Berkeley speech and later in his autobiography. See Theodore Roosevelt, "How the United States Acquired the Right to Dig the Panama Canal," *Outlook*, Oct. 7, 1911, 314–318, and Roosevelt, *Autobiography*, 516–543.

86. Marshall, *Story of the Panama Canal*, 63.

87. John Abbot, *Panama and the Canal in Picture and Prose: A Complete Story of Panama, as Well as the History, Purpose and Promise of Its World-Famous Canal—the Most Gigantic Engineering Undertaking since the Dawn of Time* (New York: Syndicate., 1913), 394.

88. Roosevelt to Taft, Dec. 6, 1910, in *Letters of Theodore Roosevelt*, 7:179.

89. Cited in Frank Morton Todd, *The Story of the Exposition: Being the Official History of the International Celebration Held at San Francisco in 1915 to Commemorate the Discovery of the Pacific Ocean and the Construction of the Panama Canal*, vol. 5 (New York: G. P. Putnam's Sons, 1921), 55.

90. Officially, there was a First Isthmian Canal Commission (sometimes referred to as the Walker Commission) constituted in 1899 to study possible canal routes; the Second Isthmian Canal Commission mentioned here covered the entire era of construction. As it is often done, I refer to the first, second, and third commissions as the ones headed by Walker, Theodore P. Shonts, and George W. Goethals, respectively, while the Canal was being built.

91. Stevens called this ubiquitous plea the "idiotic howl." Cited in Bennett, *History of the Panama Canal*, 218.

92. J. B. Bishop, *Panama Gateway*, 155.

93. F. Bishop, *Panama Past and Present*, 220.

94. Marie D. Gorgas and Burton J. Hendrick, *William Crawford Gorgas: His Life and Work* (Garden City, NY: Doubleday, Page, 1924), 153.

95. Cited in Walter Leon Pepperman, *Who Built the Panama Canal?* (London: J. M. Dent and Sons, 1915), 73.

96. Michael L. Conniff, *Black Labor on a White Canal: Panama, 1904–1981* (Pittsburgh: University of Pittsburgh Press, 1985), 3–4. As before, many of them decided to stay on the Isthmus, constituting the largest immigrant group in Panama. The resulting ethnic conflicts culminated in the deportation campaigns of the 1920s. For the Panama authors' views of the West Indians, see especially pp. 113–119 and 129–131.

97. George W. Goethals, "The Building of the Panama Canal: Labor Problems Connected with the Work," *Scribner's Magazine* 57 (April 1915), 396. On the situation of the few African Americans on the Isthmus, see Patrice C. Brown, "The Panama Canal: The African-American Experience," *Prologue: Quarterly of the National Archives and Records Administration* 29 (Summer 1997), 122–126.

98. Major, *Prize Possession*, 83.

99. Julie Greene, "Spaniards on the Silver Roll: Labor Troubles and Liminality in the Panama Canal Zone, 1904–1914," *International Labor and Working-Class History* 66 (Fall 2004): 79.

100. Harry A. Franck, *Zone Policeman 88: A Close Range Study of the Panama Canal and Its Workers* (New York: Century, 1913), 165.

101. C. A. McIlvaine, Executive Secretary, to Walter V. Eagleson, Sept. 2, 1914; Race Question in the Canal Zone, File 28-B-233, Part 1; Records of the Panama Canal 1851–1960, General Records 1914–1934, Record Group 185, National Archives at College Park, Maryland.

102. Abbot, *Panama and the Canal*, 188.

103. "The Land of the Cocoanut-Tree," in *Panama Patchwork: Poems by James Stanley Gilbert* (Colón, Panama: J. V. Beverhoudt, 1920), 1–2. One of Gilbert's favorite places was the Panamanian cemetery near Mount Hope. The poet died in 1906.

104. Roosevelt to Walker, Feb. 24, 1904, in *Letters of Theodore Roosevelt*, 4:738.

105. "Yellow Eyes," in *Panama Patchwork*, 17.

106. See McCullough, *Path between the Seas*, 447, 451, 468. On the details of the sanitation efforts and their interpretation, see pp. 56–63.

107. Major, *Prize Possession*, 69.

108. Pepperman, *Who Built the Panama Canal?* 193. Another rerouted railroad had to be built later due to the creation of Gatun Lake.

109. *Speech of Theodore P. Shonts, Chairman of the Isthmian Canal Commission, Before the American Hardware Manufacturers' Association, at the New Willard Hotel, Washington, DC, on the Evening of November 9, 1905* (n.p.), 20.

110. Poultney Bigelow, "Our Mismanagement at Panama," *Independent*, Jan. 4, 1906, 9–21. On the Bigelow affair, see also J. M. Hogan, *Panama Canal*, 39–43.

111. On the muckrakers, see also chap 4.

112. See Willis Fletcher Johnson, *Four Centuries of the Panama Canal* (New York: Cassell, 1907), 345.

113. Bigelow, "Our Mismanagement at Panama," 20.

114. See the *Independent*, Jan. 18, 1906, 127–128. On Robinson's influence, see p. 77.

115. See the *Independent*, March 15, 1906, 589–596.

116. Poultney Bigelow, "Panama—the Human Side," *Cosmopolitan Magazine* 41 (Sept. 1906), 455–462.

117. Charles Harcourt Forbes-Lindsay, *The Isthmus and the Canal*, rev. ed. (1906; Philadelphia: John C. Winston, 1912), 267.

118. Pepperman, *Who Built the Panama Canal?* 245.

119. For the following paragraphs, see also McCullough, *Path between the Seas*, 492–500, and J. M. Hogan, *Panama Canal*, 43–45.

120. "Roosevelt in Panama as Travelers Saw Him," *New York Times*, Nov. 25, 1906, 3.

121. "Address of President Roosevelt to the Employees of the Isthmian Canal Commission, at Colon, Panama, November 17, 1906," in *Special Message of the President of the United States Concerning the Panama Canal, Communicated to the Two Houses of Congress on Dec. 17, 1906* (Washington, DC: Government Printing Office, 1906), 16. Roosevelt's *Special Message* was reprinted in the *New York Times*, Dec. 18, 1906, as a supplement.

122. Mary A. Chatfield, *Light on Dark Places at Panama: By an Isthmian Stenographer* (New York: Broadway, 1908), 196.

123. "Address of President Roosevelt to the Employees," 16. Roosevelt was keenly aware of public relations (a term coined in the late nineteenth century) instruments. While his self-styled media personality may have been exceptional at the time, the use of government reports for publicity purposes had been a common practice since the 1890s, when editorial assistants were hired to prepare them. The General Printing Act of 1895 was drafted to restrict unnecessary agency publications. See Scott M. Cutlip, *Public Relations History: From the 17th to the 20th Century: The Antecedents* (Hillsdale, NJ: Lawrence Erlbaum, 1995), 223.

124. *Special Message of the President of the United States*, 4.

125. Ibid., 14.

126. Weir, *Conquest of the Isthmus*, 162, and Herbert Croly, *The Promise of American Life: With a New Introduction by Scott R. Bowman* (1909; New Brunswick, NJ: Transaction, 1993), 174.

127. For the following, see also McCullough, *Path between the Seas*, 491–492, and Major, *Prize Possession*, 69–70.

128. On Goethals's personality and rule, see pp. 142–149.

129. George W. Goethals, "The Building of the Panama Canal: Success of Government Methods," *Scribner's Magazine* 56 (March 1915), 277.

130. See Major, *Prize Possession*, 103. The second figure refers to the fiscal year of 1907.

131. See ibid., 98–101.
132. The effect of the Taft Agreement on the interpretations of the Canal Zone is discussed on p. 125.
133. See Conniff, *Panama and the United States,* 71.

Chapter 2. American Triumph

1. Ricardo D. Salvatore, "Imperial Mechanics: South America's Hemispheric Integration in the Machine Age," *American Quarterly* 58 (Sept. 2006): 665.
2. "Beyond the Chagres," *Panama Patchwork: Poems by James Stanley Gilbert* (Colón, Panama: J. V. Beverhoudt, 1920), 14.
3. Charles Francis Adams, *The Panama Canal Zone: An Epochal Event in Sanitation* (Boston: Massachusetts Historical Society, 1911), 8.
4. Ibid., 13.
5. Philip Curtin, *Death by Migration: Europe's Encounter with the Tropical World in the Nineteenth Century* (Cambridge, UK: Cambridge University Press, 1989), 159–160.
6. For the history of public health and cleanliness, see Suellen Hoy, *Chasing Dirt: The American Pursuit of Cleanliness* (Oxford: Oxford University Press, 1995), and John Duffy, *The Sanitarians: A History of American Public Health* (Urbana: University of Illinois Press, 1990).
7. Hoy, *Chasing Dirt,* 63.
8. See Hoy, *Chasing Dirt,* 89, and Richard J. Evans, *Death in Hamburg: Society and Politics in the Cholera Years, 1830–1910* (Oxford: Clarendon Press, 1987), 179.
9. See David McCullough, *The Path between the Seas: The Creation of the Panama Canal, 1870–1914* (New York: Simon and Schuster, 1977), 143. The professor suggested that the nation's capitol should be enclosed by a mosquito screen the size of the Washington Monument.
10. *Scientific American,* July 23, 1904, 58.
11. Walter Leon Pepperman, *Who Built the Panama Canal?* (London: J. M. Dent and Sons, 1915), 64.
12. See McCullough, *Path between the Seas,* 144.
13. A critical biography of Gorgas still needs to be written. The most recent study is Edward F. Dolan, *William Crawford Gorgas: Warrior in White* (New York: Dodd, Mead, 1968).
14. Marie D. Gorgas and Burton J. Hendrick, *William Crawford Gorgas: His Life and Work* (Garden City, NY: Doubleday, Page, 1924), 4.
15. William Crawford Gorgas, *Sanitation in Panama* (New York: D. Appleton, 1915), 6.
16. The name in use today is *Aedes aegypti.* The mosquito can be identified by its characteristic stripes. Only the females sting.

17. See Joseph Bucklin Bishop, *The Panama Gateway* (New York: Charles Scribner's Sons, 1913), 230.

18. See W. C. Gorgas, *Sanitation in Panama*, 42–44, 65.

19. M. Gorgas and Hendrick, *William Crawford Gorgas*, 169.

20. See William Crawford Gorgas, "Health Conditions on the Isthmus in Panama," *Scientific American Supplement*, July 16, 1904, 23856.

21. For the following paragraphs, see W. C. Gorgas, *Sanitation in Panama*, 151–153; M. Gorgas and Hendrick, *William Crawford Gorgas*, 200; and McCullough, *Path between the Seas*, 140, 419–423, 466–467. On diseases and sanitation in Panama, see David Ray Abernathy, "Bound to Succeed: Science, Territoriality and the Emergence of Disease Eradication in the Panama Canal Zone" (Ph.D. diss., University of Washington, 2000), J. P. MacLaren, *A Brief History of Sanitation in the Canal Zone, 1513–1972* (n.p., 1972), and James Steven Simmons, *Malaria in Panama* (1939; New York: Arno Press, 1979).

22. See MacLaren, *Brief History of Sanitation*, 22.

23. See J. B. Bishop, *Panama Gateway*, 248.

24. See W. C. Gorgas, *Sanitation in Panama*, 275.

25. See McCullough, *Path between the Seas*, 582.

26. Michael L. Conniff, *Black Labor on a White Canal: Panama, 1904–1981* (Pittsburgh: University of Pittsburgh Press, 1985), 31. For the official numbers, see also McCullough, *Path between the Seas*, 610.

27. See W. C. Gorgas, *Sanitation in Panama*, 156–157, and *Scientific American*, March 10, 1906, 214–215.

28. See MacLaren, *Brief History of Sanitation*, 20, 23. To adjust the total costs of $400 million to today's purchasing power, a ratio somewhere between 1:10 and 1:20 should be assumed. See Walter LaFeber, *The Panama Canal: The Crisis in Historical Perspective* (1978; Oxford: Oxford University Press, 1989), 219.

29. See Abernathy, "Bound to Succeed," 144–145.

30. See John Lindsay-Poland, *Emperors in the Jungle: The Hidden History of the U.S. in Panama* (Durham, NC: Duke University Press, 2003), 35.

31. "The Real Situation at Panama," *Independent*, Feb. 9, 1905, 310.

32. Poultney Bigelow, "Panama—the Human Side," *Cosmopolitan Magazine* 41 (Sept. 1906): 461. Americans had been using insect screens for decades and exported their habit to Cuba and Panama, while they were almost unknown in Great Britain and its tropical colonies. See Curtin, *Death by Migration*, 137.

33. John Abbot, *Panama and the Canal in Picture and Prose: A Complete Story of Panama, as Well as the History, Purpose and Promise of Its World-Famous Canal—the Most Gigantic Engineering Undertaking since the Dawn of Time* (New York: Syndicate, 1913), 259.

34. Frederic Jennings Haskin, *The Panama Canal* (Garden City, NY: Doubleday, Page, 1913), 107.

35. Abernathy, "Bound to Succeed," 173.

36. Ira E. Bennett, *History of the Panama Canal: Its Construction and Builders* (Washington, DC: Historical, 1915), 123.

37. See Abbot, *Panama and the Canal*, 260; Charles Harcourt Forbes-Lindsay, *The Isthmus and the Canal*, rev. ed. (1906; Philadelphia: John C. Winston, 1912), 279; and Pepperman, *Who Built the Panama Canal?* 144–145.

38. Goethals, Notice to Employees, Oct. 21, 1907; Methods of Storing Equipment, Material, Scrap, etc., to prevent breeding of Mosquitoes and Rats, File 37-F-35, Part I; Records of the Second Isthmian Canal Commission 1904–16, General Correspondence (GC) 1904–14, Record Group (RG) 185, National Archives at College Park (NACP), Maryland.

39. Gorgas, Notice to Managers of Hotels, Restaurants and Eating Houses in the Canal Zone, April 22, 1909; Inspections of Hotels, Messes, and Kitchens by Health Dept., File 37-F-14, GC 1904–14, RG 185, NACP.

40. Haskin, *Panama Canal*, 116. The interpretations of the Canal Zone state are discussed in chap. 4. Interestingly, Gorgas himself reports an incident of resistance against the sanitary regulations: In a sanatorium the French had built on the island of Taboga near Panama City, the patients had to take quinine as a preventive medicine against malaria. Some refused and instead fed the tablets to their house pet, a turkey. The animal then developed the so-called quinine blindness. See W. C. Gorgas, *Sanitation in Panama*, 223.

41. Joseph Bucklin Bishop, "Sanitation on the Isthmus," *Scribner's Magazine* 53 (Feb. 1913), 234.

42. Abbot, *Panama and the Canal*, 36, 254.

43. See also Abernathy, "Bound to Succeed," 201.

44. Goethals to Gorgas, May 16, 1912; Inquiries and Statements re Health Conditions and Sanitary Work on the Isthmus, File 37-E-25, part 2, GC 1904–14, RG 185, NACP.

45. See ibid.

46. W. C. Gorgas, *Sanitation in Panama*, 285. On Gorgas's vision, see also Forbes-Lindsay, *Isthmus*, 37.

47. William Joseph Showalter, "Redeeming the Tropics," *National Geographic Magazine* 25 (March 1914): 357.

48. Brenda Gayle Plummer, *Haiti and the United States: The Psychological Moment* (Athens: University of Georgia Press, 1992), 90.

49. Willis Fletcher Johnson, *Four Centuries of the Panama Canal* (New York: Cassell, 1907), 280.

50. Logan Marshall, *The Story of the Panama Canal* (Philadelphia: John C. Winston, 1913), 7.

51. See especially David E. Nye, *American Technological Sublime* (Cambridge, MA: MIT Press, 1994); Nye, *Narratives and Spaces: Technology and the Construction of American Culture* (Exeter, UK: University of Exeter Press, 1997); and Nye, *America as Second Creation: Technology and Narratives of New Beginnings* (Cambridge, MA:

MIT Press, 2003). On the earlier literature on public works, see Suellen Hoy and Michael C. Robinson, eds., *Public Works History in the United States: A Guide to the Literature* (Nashville: American Association for State and Local History, 1982).

52. Nye, *American Technological Sublime*, xx. Nye explains the attraction of these projects with their evocation of the sublime, "an essentially religious feeling" (xiii). Complementing Leo Marx's *The Machine in the Garden: Technology and the Pastoral Ideal in America* (Oxford: Oxford University, 1964), he introduces the concept of a technological sublime that encompasses the transformation of "natural" wonders such as the Grand Canyon and the Niagara Falls into man-made landscapes. For a discussion of the sublime with regard to the images of the Panama Canal, see pp. 84–88 and 106.

53. Nye, *America as Second Creation*, 150, 231.

54. "The Grand Canal Celebration," *Utica (NY) Sentinel*, Nov. 8, 1825, cited in Nye, *American Technological Sublime*, 36. On the building and meanings of the Erie Canal, see also Carol Sheriff, *The Artificial River: The Erie Canal and the Paradox of Progress, 1817–1862* (New York: Hill and Wang, 1996).

55. "The Grand Canal Celebration," cited in Nye, *American Technological Sublime*, 36.

56. Cited in David McCullough, *The Great Bridge: The Epic Story of the Building of the Brooklyn Bridge* (New York: Simon and Schuster, 1972), 536. On the Brooklyn Bridge, see also Alan Trachtenberg, *Brooklyn Bridge: Fact and Symbol* (Chicago: University of Chicago Press, 1965), and most recently Richard Haw, *The Brooklyn Bridge: A Cultural History* (New Brunswick, NJ: Rutgers University Press, 2005).

57. See Dirk Van Laak, *Weiße Elefanten: Anspruch und Scheitern technischer Großprojekte im 20. Jahrhundert* (Stuttgart: Deutsche Verlags-Anstalt, 1999), 32, and Matt K. Matsuda, *Empire of Love: Histories of France and the Pacific* (Oxford: Oxford University Press, 2005), 51–53. Barthélemy-Prosper Enfantin, a follower of the aristocrat Claude-Henri de Saint-Simon (1760–1825), started a canal project in Egypt, wishing to unite the East and the West, but then had to abandon it. For his own project, de Lesseps drew on the plans of the Saint-Simonians. See Zachary Karabell, *Parting the Desert: The Creation of the Suez Canal* (London: John Murray, 2003), 71.

58. Frank A. Gause and Charles Carl Carr, *The Story of Panama: The New Route to India* (Boston: Silver, Burdett, 1912), 159.

59. J. B. Bishop, *Panama Gateway*, 3.

60. Johnson, *Four Centuries of the Panama Canal*, 388.

61. Cited in John Logan Allen, *Passage through the Garden: Lewis and Clark and the Image of the American Northwest* (Urbana: University of Illinois Press, 1975), xix.

62. See Allen, *Passage through the Garden*, 110–111.

63. See Henry Nash Smith, *Virgin Land: The American West as Symbol and Myth* (1950; Cambridge, MA: Harvard University Press, 1978), 22–34.

64. Harry Harwood Rousseau, *The Isthmian Canal: Presented at the Twentieth Annual Session of the Trans-Mississippi Commercial Congress Held at Denver, Colo., August 16–21, 1909* (Washington, DC: Government Printing Office, 1910), 3.

65. Walt Whitman, *Leaves of Grass: The 1892 Edition* (1892; New York: Bantam Books, 1983), 329.

66. Ralph Emmett Avery, *America's Triumph at Panama: Panorama and Story of the Construction and Operation of the World's Giant Waterway from Ocean to Ocean* (Chicago: Regan Printing House, 1913), 372.

67. Johnson, *Four Centuries of the Panama Canal,* 388.

68. Abbot, *Panama and the Canal,* 61.

69. Bill Brown, "Science Fiction, the World's Fair, and the Prosthetics of Empire, 1910–1915," in *Cultures of United States Imperialism,* ed. Amy Kaplan and Donald E. Pease (Durham, NC: Duke University Press, 1993), 148.

70. Ben Macomber, *The Jewel City: Its Planning and Achievement; Its Architecture, Sculpture, Symbolism, and Music; Its Gardens, Palaces, and Exhibits* (San Francisco: John H. Williams, 1915), 11.

71. Stephen Kern, *The Culture of Time and Space, 1880–1918* (Cambridge, MA: Harvard University Press, 1983), 240.

72. Alfred Thayer Mahan, *The Interest of America in Sea Power: Present and Future* (1897; Port Washington, NY: Kennikat Press, 1970), 124.

73. Cited in *History of the Panama-Pacific International Exposition, Comprising the History of the Panama Canal and a Full Account of the World's Greatest Exposition, Embracing the Participation of the States and Nations of the World and other Events at San Francisco, 1915, Compiled by the Pan-Pacific Press Association, Ltd.* (n.d.), 7. The Italian Guglielmo Marconi (1874–1937) had invented wireless telegraphy in 1896.

74. An interesting attempt to draw parallels between contemporary and earlier communication technologies is Tom Standage, *The Victorian Internet: The Remarkable Story of the Telegraph and the Nineteenth Century's On-Line Pioneers* (New York: Walker, 1998).

75. Matsuda, *Empire of Love,* 47.

76. Marshall, *Story of the Panama Canal,* 7.

77. Michael Hardt and Antonio Negri, *Empire* (Cambridge, MA: Harvard University Press, 2000), xiv.

78. J. B. Bishop, *Panama Gateway,* 194.

79. After the completion of the Canal, the cut was renamed as Gaillard Cut as a tribute to Colonel David DuBose Gaillard, who was in charge of the work until his death in 1913. See the *Canal Record,* May 12, 1915, 338.

80. Haskin, *Panama Canal,* 71.

81. John Foster Fraser, *Panama and What It Means* (London: Cassell, 1913), 39.

82. Avery, *America's Triumph,* 119.

83. An icon of the seal also appeared on a bronze medal, coined in 1908 in Philadelphia. This was the badge of honor Roosevelt had promised the Canal

laborers on his visit to the zone. The other side showed an image of TR. See F. Bishop, *Panama Past and Present*, 248; and William Rufus Scott, *The Americans in Panama* (New York: Statler, 1912), 220.

84. See McCullough, *Path between the Seas*, 543.

85. Arthur Bullard [Albert Edwards], *Panama: The Canal, the Country and the People*, rev. ed. with additional chapters (1911; New York: Macmillan, 1914), 47.

86. *Photogravure Reproductions of the Panama Canal* (Passaic, NJ: Rotary Photogravure, 1913), n.p.

87. *Canal Record*, Aug. 13, 1913, 429. On the history of tourism, see John F. Sears, *Sacred Places: American Tourist Attractions in the Nineteenth Century* (1989; Amherst: University of Massachusetts Press, 1998), and Marguerite S. Shaffer, *See America First: Tourism and National Identity, 1880–1940* (Washington, DC: Smithsonian Institution Press, 2001). On the tourist perspective of the Canal Zone, see also pp. 126–127.

88. *Canal Record*, Dec. 18, 1912, 133.

89. "Tourists and Touristesses: An Interview with Wm. M. Baxter, Official Guide," in *Society of the Chagres, Year Book 1913* (Culebra, CZ: John O. Collins, n.d.), 60–61.

90. Ibid., 63.

91. Avery, *America's Triumph*, 7.

92. *Scientific American*, Nov. 9, 1912, 391. Abbot, *Panama and the Canal*, 134, reprinted the graphic comparison. See also McCullough, *Path between the Seas*, 529, who notes that the analogies were "seldom any less fantastic" than the Canal statistics.

93. Scott, *Americans at Panama*, xi–xii.

94. See, for instance, C. F. Adams, *Panama Canal Zone*, 13.

95. A. W. Wyndham, *The Panama Canal* (New York: Howard F. Curtis, 1907), 25.

96. Bernhard Kellermann, *Der Tunnel* (Berlin: S. Fischer, 1913). The English translation was published in 1915.

97. The Sam deVincent Collection of Illustrated American Sheet Music at the National Museum of American History includes five songs and three marches. I would like to thank Marcia Rodwin, Archives Center, for providing this information.

98. See Janet M. Davis, *The Circus Age: Culture and Society under the American Big Top* (Chapel Hill: University of North Carolina Press, 2002), 210. I would like to thank the author for the reference.

99. *Town and Country*, May 16, 1914, 56; also reproduced in Susan Strasser, *Satisfaction Guaranteed: The Making of the American Mass Market* (New York: Pantheon Books, 1989), 110. It was calculated that a twelve-thousand-ton cargo ship would save twenty-six days of its journey (and more than $3,800 in costs after tolls) from New York to San Francisco by using the Canal. See *Official*

Handbook of the Panama Canal 1915 (Washington, DC: Government Printing Office, 1915), 32.

100. Hugh C. Weir, *The Conquest of the Isthmus: The Men Who Are Building the Panama Canal—Their Daily Lives, Perils, and Adventures* (New York: G. P. Putnam's Sons, 1909), 86.

101. Logan Marshall, *Seeing America: Including the Panama Expositions* (Philadelphia: John C. Winston, 1915), 129.

102. Abbot, *Panama and the Canal*, 36, 159.

103. "Address of President Roosevelt to the Employees," in *Special Message of the President of the United States Concerning the Panama Canal, Communicated to the Two Houses of Congress on Dec. 17, 1906*, 16.

104. *Scientific American*, Nov. 11, 1912, 384.

105. Cited in J. B. Bishop, *Panama Gateway*, 178.

106. The fortification of the Canal, part of Mahan's concept of "military preparedness," was legal according to the second version of the Hay-Pauncefote Treaty drafted in 1902, but during the construction era it was repeatedly contested by critics who stressed the role of the Canal in promoting global peace. Only some coastal batteries were installed, and the army sent an understaffed infantry regiment to the Isthmus in 1911. See John Major, *Prize Possession: The United States and the Panama Canal, 1903–1979* (Cambridge, UK: Cambridge University Press, 1993), 156–161. During World War I, in light of German submarine campaigns, better fortification once again seemed imperative. In his novel *The Conquest of America: A Romance of Disaster and Victory, U.S.A., 1921 A.D, Based on Extracts from the Diary of James E. Langston, War Correspondent of the "London Times"* (New York: George H. Doran, 1916), Cleveland Moffett, a proponent of the "two-ocean navy," imagined a German suicide commando entering the Canal on a merchant vessel carrying six hundred tons of dynamite designated to destroy the Canal and pave the way for a German invasion of the United States.

107. Cited in J. B. Bishop, *Panama Gateway*, 64.

108. See Hélène Christol, "The Pacifist Warrior: William James and His 'Moral Equivalent of War,'" in *An American Empire: Expansionist Cultures and Policies, 1881–1917*, ed. Serge Ricard (Aix-en-Provence: Publications de l'Université de Provence Aix-Marseille I, 1990), 193.

109. William James, *The Moral Equivalent of War and Other Essays and Selections from Some Problems of Philosophy, Edited and with an Introduction by John K. Roth* (New York: Harper and Row, 1971), 12.

110. Ibid., 13.

111. See Olaf Stieglitz, *100 Percent American Boys: Disziplinierungsdiskurse und Ideologie im Civilian Conservation Corps, 1933–1942* (Stuttgart: Franz Steiner Verlag, 1999), 90–91, and Christopher Lasch, *The True and Only Heaven: Progress and Its Critics* (New York: W. W. Norton, 1991), 301.

112. See Christol in Ricard, *American Empire*, 191. On James's conceptions of gender, see Athena B. Devlin, "Between Profits and Primitivism: Rehabilitating White Middle-Class Manhood in America, 1880–1917" (Ph.D. diss., University of Massachusetts at Amherst, 2001); and Kim Townsend, *Manhood at Harvard: William James and Others* (New York: W. W. Norton, 1996).

113. Devlin, "Between Profits and Primitivism," 190. The embedded quotation is cited from William James, *The Varieties of Religious Experience: A Study in Human Nature* (1902; New York: Collier Books, 1961), 289.

114. J. B. Bishop, *Panama Gateway*, 63–64.

115. *Scientific American*, July 23, 1904, 58.

116. J. B. Bishop, *Panama Gateway*, 65.

117. Weir, *Conquest of the Isthmus*, 66. The tools were probably intended to remove ash from the boilers of the steam shovels. See McCullough, *Path between the Seas*, 149. Weir also hugely exaggerated the figures—the mean temperature in Culebra in 1911 was 79.2 degrees. See the *Canal Record*, Jan. 17, 1912, 169.

118. J. B. Bishop, *Panama Gateway*, 88.

119. Haskin, *Panama Canal*, 9.

120. Abbot, *Panama and the Canal*, 120.

121. C. F. Adams, *Panama Canal Zone*, 19.

122. Scott, *Americans in Panama*, 45.

123. George W. Goethals, "The Building of the Panama Canal: The Human Element in Administration," *Scribner's Magazine* 57 (June 1915): 721.

124. Goethals made this statement at the Panama-Pacific International Exposition in San Francisco, cited in Frank Morton Todd, *The Story of the Exposition: Being the Official History of the International Celebration Held at San Francisco in 1915 to Commemorate the Discovery of the Pacific Ocean and the Construction of the Panama Canal*, vol. 3 (New York: G. P. Putnam's Sons, 1921), 133.

125. W. C. Gorgas, *Sanitation in Panama*, 144.

126. Pepperman, *Who Built the Panama Canal?* 41–42.

127. See McCullough, *Path between the Seas*, 146.

128. Forbes-Lindsay, *Isthmus*, 11, and J. B. Bishop, *Panama Gateway*, 90, for example, cite Froude. Avery, *America's Triumph*, 51, and again J. B. Bishop, *Panama Gateway*, 74, rely on Robinson.

129. Tracy Robinson, *Panama: A Personal Record of Forty-Six Years, 1861–1907* (New York: Star and Herald, 1907), 239, 143.

130. See John Higham, "The Reorientation of American Culture in the 1890s," *Writing American History: Essays on Modern Scholarship* (Bloomington: Indiana University Press, 1970), 78–79, and Gail Bederman, *Manliness and Civilization: A Cultural History of Gender and Race in the United States, 1880–1917* (Chicago: University of Chicago Press, 1995), 17.

131. Joseph Bucklin and Farnham Bishop, *Goethals: Genius of the Panama Canal* (New York: Harper and Brothers, 1930), 133, cited in McCullough, *Path between the Seas*, 464.

132. Abbot, *Panama and the Canal*, 121.
133. Haskin, *Panama Canal*, 10.
134. See Matsuda, *Empire of Love*, 57.
135. Ibid., 45. About twenty-one thousand Frenchmen (of a total twenty-three thousand investors) had bought stocks.
136. Hardt and Negri, *Empire*, xv.

Chapter 3. The Engineered View

1. Ira E. Bennett, *History of the Panama Canal: Its Construction and Builders* (Washington, DC: Historical, 1915), 419.

2. A few collections of Canal images have been published. The most notable ones are Ulrich Keller, *The Building of the Panama Canal in Historic Photographs* (New York: Dover, 1983), which reproduces Ernest Hallen's official views and includes an introduction, and Jerome D. Laval, *Images of an Age: Panama and the Building of the Canal, Photographs from the Keystone-Mast Stereograph Collection* (Fresno, CA: Graphic Technology, 1978), which assembles photos first published as stereographs. Two historical publications of images are also worth mentioning: *King's Views of the Panama Canal in Course of Construction* (New York: Moses King, 1912) and the above-cited *Photogravure Reproductions of the Panama Canal* (Passaic, NJ: Rotary Photogravure, 1913). These works include short captions but no credits. An example for tainted "color" photography is Earle Harrison, *The Panama Canal: Illustrated by Color Photography from the Original Autochrome Photographs* (New York: Moffat, Yard, 1913).

3. Oliver Wendell Holmes, "The Stereoscope and the Stereograph" (1859), reprinted in *Classic Essays on Photography*, ed. Alan Trachtenberg (New Haven, CT: Leete's Island Books, 1980), 71–82. The essay was originally published in the *Atlantic Monthly*. On the use of images in historical scholarship, see the excellent overview by Jens Jäger, *Photographie: Bilder der Neuzeit, Einführung in die Historische Bildforschung* (Tübingen: edition diskord, 2000).

4. Early on in the history of American photography, the Pictorialist photographers, the most prominent of whom were Alfred Stieglitz (1864–1946) and Edward Steichen (1879–1973), had facilitated this classification. They employed "painterly" and impressionist techniques in their photographs and considered themselves artists. At the same time, these images were complex interpretations of urban industrialization.

5. Rosalind Krauss, "Photography's Discursive Spaces" (1982), reprinted in *The Contest of Meaning: Critical Histories of Photography*, ed. Richard Bolton (Cambridge, MA: MIT Press, 1989), 287–301, esp. 290–291.

6. Julie K. Brown, *Contesting Images: Photography and the World's Columbian Exposition* (Tucson: University of Arizona Press, 1994), xiii.

7. See Erwin Panofsky, *Studies in Iconology: Humanistic Themes in the Art of the Renaissance* (New York: Oxford University Press, 1939). For a modern study

based on his ideas, see Martin Warnke, *Politische Landschaft: Zur Kunstgeschichte der Natur* (München: Hanser, 1992). Alan Trachtenberg, in *Reading American Photographs: Images as History, Matthew Brady to Walker Evans* (New York: Hill and Wang, 1989), employs related methods to analyze American photography. For a critique, see Jäger, *Photographie*, 75–76. Typical strategies of Panofsky's approach, such as the analysis of allegories, cannot be applied to photography.

8. On this approach, see Krauss, "Photography's Discursive Spaces," esp. 295–299, as well as Allan Sekula, "The Body and the Archive," in Bolton, *Contest of Meaning*, 343–389, and John Tagg, *The Burden of Representation: Essays on Photographies and Histories* (Minneapolis: University of Minnesota Press, 1988).

9. In his conversational essay "The Eye of Power," Foucault employs a metaphor, the panopticon, to illustrate the process of observation. This optical device enables the overseer of a prison to observe every movement of the inmates. The transparency and ubiquity of control has replaced the protective darkness of the dungeon. To illustrate that the discursive machinery is of greater relevance than its operators, Foucault asks his interviewer the famous rhetorical question: "Do you think it would be much better to have the prisoners operating the Panoptic apparatus and sitting in the central tower, instead of the guards?" Michel Foucault, "The Eye of Power," in *Power/Knowledge: Selected Interviews and Other Writings, 1972–1977,* ed. Colin Gordon (New York: Harvester Wheatsheaf, 1980), 165.

10. Krauss, "Photography's Discursive Spaces," 293.

11. See F. Jack Hurley, *Industry and the Photographic Image: 153 Great Prints from 1850 to the Present* (New York: Dover, 1980), 44.

12. Cited in *Encyclopedia of Photography,* ed. International Center of Photography (New York: Crown, 1984), 556.

13. See Michael L. Carlebach, *American Photojournalism Comes of Age* (Washington, DC: Smithsonian Institution Press, 1997), 19. Jäger, *Photographie*, 61, criticizes the "myth" that the Kodak quickly became an actual mass product. For an overview of the technological and economic development of early photography, see Reese V. Jenkins, *Images and Enterprise: Technology and the American Photographic Industry, 1839 to 1925* (Baltimore: Johns Hopkins University Press, 1975).

14. See *Encyclopedia of Photography,* 294.

15. See William Welling, *Photography in America: The Formative Years, 1839–1900* (New York: Thomas Y. Crowell, 1978), 313. For the impact of the halftone effect, see also Neil Harris, *Cultural Excursions: Marketing Appetites and Cultural Tastes in Modern America* (Chicago: University of Chicago Press, 1990), 307.

16. See Carlebach, *American Photojournalism Comes of Age,* 3.

17. See ibid., 65.

18. See ibid., 32.

19. Jürgen Martschukat, "The Art of Killing by Electricity: The Sublime and the Electric Chair," *Journal of American History* 89 (Dec. 2002), 903–904. For

an extended analysis see David E. Nye, *American Technological Sublime* (Cambridge, MA: MIT Press, 1994); and Nicholas Mirzoeff, *An Introduction to Visual Culture* (London: Routledge, 1999), 16.

20. John Foster Fraser, *Panama and What It Means* (London: Cassell, 1913), 39.

21. Albert Boime, *The Magisterial Gaze* (Washington, DC: Smithsonian Institution Press, 1991), n.p.

22. Nye, *American Technological Sublime*, xiv.

23. For an analysis of survey photography, see Trachtenberg, *Reading American Photographs*, esp. the introduction and 127–129. For its promotional value, see Nancy K. Anderson, "'The Kiss of Enterprise': The Western Landscape as Symbol and Resource," in *The West as America: Reinterpreting Images of the Frontier, 1820–1920*, ed. William H. Truettner (Washington, DC: Smithsonian Press, 1991), 256. For a broader comparative approach, see Joan M. Schwartz and James R. Ryan, eds., *Picturing Place: Photography and the Geographical Imagination* (London: I. B. Tauris, 2003).

24. See Martin W. Sandler, *American Image: Photographing One Hundred Fifty Years in the Life of a Nation* (Chicago: Contemporary Books, 1989), 56. Many of the photographic survey records are held at the National Archives in College Park, Maryland. See *The American Image: Photographs from the National Archives, 1860–1960* (New York: Pantheon Books, 1979).

25. The archive of the company, later renamed Detroit Publishing Company, can be searched through the Library of Congress's American Memory digital collection, http://memory.loc.gov/ammem/detroit/dethome.html, accessed Sept. 22, 2007.

26. Cited in Martha A. Sandweiss, "Undecisive Moments: The Narrative Tradition in Western Photography," in *Photography in Nineteenth Century America* (New York: Harry N. Abrams, 1991), 119. Sandweiss shows that depending on the objectives of the expedition leader, O'Sullivan's pictures served to convey different messages: whereas King was appalled by the forces of nature, Wheeler identified a landscape ready for human exploitation through settlement and tourism.

27. Mary Louise Pratt, *Imperial Eyes: Travel Writing and Transculturation* (London: Routledge, 1992), 204–205.

28. For an account of the expedition, see David McCullough, *Path between the Seas: The Creation of the Panama Canal, 1870–1914* (New York: Simon and Schuster, 1977), 19–22, 40–44.

29. These photos are located in Lot 13472 (H), Prints and Photographs Division, Library of Congress, Washington, DC. Another government photographer who produced images of the Isthmus was Eadweard Muybridge, who became famous for his motion studies and city photography. See Lot 12067 (S), Prints and Photographs Division, Library of Congress.

30. The rock formation was located on Jefferson's property in Rockbridge County. He termed it "the most sublime of nature's works." See Thomas Jefferson, "Notes on the State of Virginia," in *The Portable Thomas Jefferson*, ed. Merrill D. Peterson (New York: Penguin Books, 1975), 54.

31. William Culp Darrah, *The World of Stereographs* (Gettysburg, PA: W. C. Darrah, 1977), 141.

32. See Hurley, *Industry and the Photographic Image*, 3.

33. See Jacob A. Riis, *Theodore Roosevelt the Citizen* (New York: Outlook, 1904).

34. See Sekula, "The Body and the Archive," 374. The standard work on documentary photographers is Maren Stange, *Symbols of the Ideal Life: Social Documentary Photography in America, 1890–1950* (Cambridge, UK: Cambridge University Press, 1989).

35. Very little work on stereographs has been published. For valuable discussions, see Edward W. Earle, ed., *Points of View: The Stereograph in America—A Cultural History* (Rochester, NY: Visual Studies Workshop, 1979), and Darrah, *World of Stereographs*. Also useful (as an inventory) is John Waldsmith, *Stereo Views: An Illustrated History and Price Guide*, 2nd ed. (Iola, WI: Krause, 2002).

36. Holmes, "Stereoscope and the Stereograph," 77.

37. Sears, Roebuck and Co., for instance, sold a set of sixty cards of the San Francisco earthquake for 75 cents.

38. Krauss, "Photography's Discursive Spaces," 290.

39. See Thomas Southall, "White Mountain Stereographs and the Development of a Collective Vision," in Earle, *Points of View*, 104–105, and Krauss, "Photography's Discursive Spaces," 291.

40. Edward W. Earle, "The Stereograph in America: Pictorial Antecedents and Cultural Perspectives," in Earle, *Points of View*, 19–20. For a discussion of colonial stereographs, see also Jim Zwick, ed., *Stereoscopic Visions of War and Empire*, originally published at http://www.boondocksnet.com/stereo/colonies.html, which was shut down in Aug. 2007.

41. See Waldsmith, *Stereo Views*, 5.

42. Such images can sometimes be identified by their negative number: V stands for Underwood, W for H. C. White. See Darrah, *World of Stereographs*, 49. Keystone continued to produce stereographs until the 1960s, when it refocused its business activities on the production of eye-testing equipment. The company was eventually bought by Mast Industries, which donated the stereograph collection—now called the Keystone-Mast Collection—to the University of California at Riverside. See http://138.23.124.165/, accessed Sept. 22, 2007.

43. Peter Bacon Hales, "American Views and the Romance of Modernization," in Sandweiss, *Photography in Nineteenth Century America*, 250.

44. Earle, "Stereograph in America," 19.

45. See Krauss, "Photography's Discursive Spaces," 290.

46. See Darrah, *World of Stereographs*, 141. Waldsmith, *Stereo Views*, 205, lists two additional twenty-four-card and thirty-five-card sets published by Keystone.

47. On postcards, see George and Dorothy Miller, *Picture Postcards in the United States, 1893–1918* (New York: Clarkson N. Potter, 1976), and Christraud M. Geary and Virginia-Lee Webb, eds., *Delivering Views: Distant Cultures in Early Postcards* (Washington, DC: Smithsonian Institution Press, 1998).

48. See Enrico Sturani, "Das Fremde im Bild: Überlegungen zur historischen Lektüre kolonialer Postkarten," *Fotogeschichte* 21, no. 79 (2001): 20.

49. Carlebach, *American Photojournalism Comes of Age*, 26.

50. Cited in Hugh C. Weir, *The Conquest of the Isthmus: The Men Who Are Building the Panama Canal—Their Daily Lives, Perils, and Adventures* (New York: G. P. Putnam's Sons, 1909), 155.

51. "The President Climbs a Canal Steam Shovel," *New York Times*, Nov. 17, 1906, p. 1.

52. Version A was printed in *Harper's Weekly*, Dec. 8, 1906, p. 1741, as part of the article by William Inglis, "At Double-Quick along the Canal with the President," 1740–1745, which included further photos of his visit credited to news syndicates. For the title page, the editors used another image credited to H. C. White that puts TR's figure further to the left of the center. This version C resembles a photograph credited to Brown Bros. that was printed in the *New York Times*, Nov. 25, 1906, part 3: Magazine Section, n.p. Version B, for instance, appeared in John Abbot's immensely popular *Panama and the Canal in Picture and Prose: A Complete Story of Panama, as Well as the History, Purpose and Promise of Its World-Famous Canal—the Most Gigantic Engineering Undertaking since the Dawn of Time* (New York: Syndicate, 1913), 125. Version A is also reproduced in U. Keller, *Building of the Panama Canal*, no. 20, version B in Laval, no. 50.

53. Additional images in the *Harper's Weekly* issue, Dec. 8, 1906, e.g., page 1745, show that most other dignitaries were wearing dark coats and suits.

54. Bill Brown, "Science Fiction, the World's Fair, and the Prosthetics of Empire, 1910–1915," in *Cultures of United States Imperialism*, ed. Amy Kaplan and Donald E. Pease (Durham, NC: Duke University Press, 1993), 139.

55. *Joseph Pennell's Pictures of the Panama Canal: Reproductions of a Series of Lithographs Made by Him on the Isthmus of Panama, January–March, 1912, Together with Impressions and Notes by the Artist* (Philadelphia: J. B. Lippincott, 1912), xiv. Interestingly, another author assigned female traits to the shovel: "She is not exactly good-looking, but mighty amiable. She grumbles considerably, and sometimes grunts and snorts in an unladylike way. But the steam-shovel man, knowing her whims, pets her a bit and says 'Please,' and up she comes with a load that fills a quarter of a flat car!" Arthur Bullard [Albert Edwards], *Panama: The Canal, the Country and the People*, rev. ed. with additional chapters (1911; New York: Macmillan, 1914), 550.

56. See *Special Message of the President of the United States Concerning the Panama Canal, Communicated to the Two Houses of Congress on December 17, 1906* (Washington, DC: Government Printing Office, 1906), which was reprinted in the *New York Times,* Dec. 18, 1906, as a supplement.

57. Joseph Bucklin Bishop, *Theodore Roosevelt and His Time: Shown in His Own Letters,* vol. 1 (New York: Charles Scribner's Sons, 1920), 454. The interchange in the Senate is retold in the article "Mr. Roosevelt's Panama Illustrations Will Not Go in The Record," *New York Times,* Dec. 18, 1906, p. 1. Vice President Charles W. Fairbanks's statement that Roosevelt's message "be printed" apparently caused confusion among the senators. The argument resulted in the agreement that "what has been read will go into The Record," but not the pictures. This outcome was greeted with "an audible snicker."

58. J. B. Bishop, *Theodore Roosevelt and His Time,* 1:455.

59. J. F. Wallace, Chief Engineer, to J. G. Walker, Isthmian Canal Commission Chairman, Feb. 23, 1905; Sale and Distribution of Photographs and Blueprints—General, File 56-A-16, part 1, General Correspondence (GC) 1904–14, Record Group (RG) 185, National Archives at College Park (NACP), Maryland. Again, the term "cold feet" is used to denote cowardly, unmanly behavior.

60. See Franz Biedermann card; Service Record Cards for Personnel Employed During Years 1904–20, RG 185, NACP.

61. F. Biedermann, Official Photographer, to E. P. Shannon, Secretary to the Chief Engineer, Sept. 11, 1905; Photographic Supplies and Equipment for Official Photographer, File 56-A-1, part 1, GC 1904–14, RG 185, NACP.

62. J. F. Stevens, Chief Engineer, to T. P. Shonts, Isthmian Canal Commission Chairman, Sept. 4, 1905; Official Photographs; Policy, Records, Instructions to Photographers, Etc.—General, File 56-A-7, part 1, GC 1904–14, RG 185, NACP.

63. James R. Mann, Member of Congress, to William H. Taft, Secretary of War, May 24, 1906, in ibid.

64. See Stevens to Shonts, June 1, 1906, in ibid.

65. Stevens to Shonts, Oct. 29, 1906, in ibid.

66. See Memorandum of Accounts Vouchered in Favor of E. Hallen, Photographer, May 22, 1907, in ibid.

67. Edward J. Williams, Disbursing Officer, to Robert M. Sands, Chief Clerk to Major Sibert, May 22, 1907, in ibid.

68. F. B. Maltby, Division Engineer, to Edward J. Williams, Disbursing Officer, April 26, 1907, in ibid.

69. W. L. Sibert, Supervisory Engineer, Memorandum for the Chief Engineer, June 15, 1907, in ibid.

70. See Stevens to J. G. Holcombe, Division Engineer, Aug. 30, 1906, and F. B. Maltby, Principal Assistant Engineer, to E. P. Shannon, Secretary to Chief Engineer, Feb. 7, 1907, in ibid.

71. See Ernest Hallen card; Service Record Cards for Personnel Employed during Years 1904–20, RG 185, NACP.

72. Ernest Hallen, *The New Pacific Fleet: Through the Panama Canal, July 1919* (Newark, NJ: Panama Pictorial, 1919). Basic facts on Hallen are provided by William Friar, *Portrait of the Panama Canal: From Construction to the Twentieth Century*, rev. and updated ed. (Portland, OR: Graphic Arts Center, 1999), 4, 31, as well as James L. Shaw, *Ships of the Panama Canal* (Annapolis, MD: Naval Institute Press, 1985), 1, 8.

73. See List of Nonexpendable Property in the Photographic Studio, Nov. 6, 1906; File 56-A-1, part 1, Photographic Supplies and Equipment for Official Photographer, GC 1904–14, RG 185, NACP.

74. A. B. Nichols, Office Engineer, to Goethals, Sept. 18, 1907; Sale and Distribution of Photographs and Blueprints—General, File 56-A-16, part 1, GC 1904–14, RG 185, NACP.

75. Goethals, Instructions in Regard to Filing Photographs and Sale of Same to Employees, n.d., in ibid.

76. Joseph Bucklin Bishop, Secretary, Isthmian Canal Commission, to Goethals, Aug. 19, 1907; Official Photographs; Policy, Records, Instructions to Photographers, Etc.—General, File 56-A-7, part 1, GC 1904–14, RG 185, NACP.

77. *Canal Record*, Dec. 12, 1908, 115.

78. See H. H. Rousseau, Assistant to Chief Engineer, to Goethals, Feb. 24, 1913; Sale and Distribution of Photographs and Blueprints—General, File 56-A-16, part 1, GC 1904–14, RG 185, NACP; and Goethals to Major F. C. Boggs, Chief of Washington Office, March 1, 1913, in ibid.

79. See *Canal Record*, Feb. 2, 1908, 190.

80. Boggs to Goethals, June 27, 1913; Sale and Distribution of Photographs and Blueprints—General, File 56-A-16, part 1, GC 1904–14, RG 185, NACP.

81. "The sale of photographs to the public was discontinued because orders had become so numerous that the official photographer was unable to fill them and at the same time attend to his other duties." Goethals to Boggs, April 17, 1911, in ibid.

82. *Canal Record*, March 29, 1911, 247.

83. See *Canal Record*, April 29, 1914, 345.

84. The correspondence can be found in Lantern Slides Illustrative of the Work on the Isthmus and Lectures Involving the Use of Lantern Slides, File 56-A-36, part 1, GC 1904–14, RG 185, NACP.

85. University of the State of New York, Educational Department Bulletin, Slides and Photographs, Panama Canal Zone and Vicinity, Dec. 1, 1912; Loans of Negatives, Photographic Plates, Lantern Slides, Moving Picture Films, Etc. to Outsiders, File 56-A-4, GC 1904–14, RG 185, NACP.

86. See, for example, College of Mechanical and Electrical Engineering, University of North Dakota, to Boggs, Feb. 4, 1914; Taking and Manufacturing

of Motion Pictures for Canal Advertisement and Publicity Purposes, File 56-C-14, part 1, GC 1904–14, RG 185, NACP.

87. See Photographs of the Construction of the Panama Canal, 1887–1940, RG 185-G, NACP. A microfilm copy of these negatives as well as prints of more than 2,200 photos were moved from the Isthmus to the National Archives in 1967. See a memorandum regarding the Photographic Records Covering the Construction Period of the Panama Canal, May 9, 1967, which is available in the NACP Still Picture Unit upon request. According to this document, it is unclear which criteria were used in selecting the 2,200 images. It only indicates that Canal engineers, members of the Canal Historical Society, and local librarians participated in the process. It is also unclear at what time between 1939 and 1967 these prints were picked and bound into twenty-four volumes. The negatives were transferred in two accretions in 1978 and 1990. In the respective finding aid, a figure of 12,000 (not 16,000) negatives is given. The discussion of images in this study is largely based on the available prints. The photos on microfilm can be accessed through an index, but they are not arranged in any chronological or topical order. A quantitative study of these images would probably provide further insights into Hallen's work.

88. For instance, Ulrich Keller states that Hallen arrived on the Isthmus on Aug. 7, 1907 (*Building of the Panama Canal*, ix), and that his predecessor's identity "may always remain a secret" (x).

89. U. Keller, *Building of the Panama Canal*, x.

90. See Elliott Woods, Superintendent U.S. Capitol Building and Grounds, to Goethals, Feb. 21, 1913; Taking and Manufacturing of Motion Pictures for Canal Advertisement and Publicity Purposes, File 56-C-14, part 1, GC 1904–14, RG 185, NACP.

91. See Woods to Goethals, April 19, 1913, in ibid.

92. Pathe Freres Company to Woods, Dec. 2, 1913, in ibid.

93. See, for example, the Precision Machine Co., Inc., to Goethals, Oct. 26, 1914; Taking and Manufacturing of Motion Pictures for Canal Advertisement and Publicity Purposes, File 56-C-14, part 2, General Records 1914–1934, RG 185, NACP. The company offered its products and services for moving picture exhibits of the Canal at the World's Fair. See also p. 183.

94. The major publications examined were John Abbot, *Panama and the Canal;* Ralph Emmett Avery, *America's Triumph at Panama: Panorama and Story of the Construction and Operation of the World's Giant Waterway from Ocean to Ocean* (Chicago: Regan Printing House, 1913); Bennett, *History of the Panama Canal;* Farnham Bishop, *Panama Past and Present* (New York: Century, 1913); Joseph Bucklin Bishop, *The Panama Gateway* (New York: Charles Scribner's Sons, 1913); Bullard, *Panama;* Charles Harcourt Forbes-Lindsay, *Isthmus and the Canal*, rev. ed. (1906; Philadelphia: John C. Winston, 1912); Harry A. Franck, *Zone Policeman 88: A Close Range Study of the Panama Canal and Its Workers* (New York: Century, 1913);

Frank A. Gause and Charles Carl Carr, *The Story of Panama: The New Route to India* (Boston: Silver, Burdett, 1912); Frederic Jennings Haskin, *The Panama Canal* (Garden City, NY: Doubleday, Page, 1913); Willis Fletcher Johnson, *Four Centuries of the Panama Canal* (New York: Cassell, 1907); Logan Marshall, *The Story of the Panama Canal* (Philadelphia: John C. Winston, 1913); William Rufus Scott, *The Americans in Panama* (New York: Statler, 1912); and Weir, *Conquest of the Isthmus*. Forbes-Lindsay used different illustrations than Marshall, even though the texts are largely identical. His and Johnson's books are based on manuscripts written in 1906 and 1907, which may be the reason why they were not yet able to use Hallen's photos. Among others, Forbes-Lindsay relied on the photographer W. A. Fishbaugh, and Johnson used images provided by news syndicates.

95. Reproduced in Bennett, *History of the Panama Canal*, 212, and Gause and Carr, *Story of Panama*, 119.

96. Reproduced in Abbot, *Panama and the Canal*, 257, and Bullard, *Panama*, 400.

97. Reproduced in Gause and Carr, *Story of Panama*, 23–24, and Alfred B. Hall and Clarence L. Chester, *Panama and the Canal* (New York: Newson, 1910), 141, 143.

98. Michael Adas, *Dominance by Design: Technological Imperatives and America's Civilizing Mission* (Cambridge, MA: Belknap Press of Harvard University Press, 2006), 169.

99. Avery, *America's Triumph*, 348.

100. U. Keller, *Building of the Panama Canal*, x.

101. J. B. Bishop, *Panama Gateway*, 195.

102. *Joseph Pennell's Pictures*. See J. Michael Hogan, *The Panama Canal in American Politics: Domestic Advocacy and the Evolution of Policy* (Carbondale: Southern Illinois University Press, 1986), 46; and Alfred Charles Richard Jr., *The Panama Canal in American National Consciousness, 1870–1990* (New York: Garland, 1990), 217.

103. See J. B. Bishop, *Panama Gateway*, 313.

104. In 1910, Pennell and fellow artists founded the Senefelder Club in London, named after the inventor of lithography, to promote their techniques.

105. In 1915, at the request of Chief Engineer Goethals, New York artist William B. Van Ingen created a group of murals that was shipped to the Canal Zone and installed in the rotunda of the Administration Building in Balboa Heights. Van Ingen's depictions of the Canal construction are very similar to Pennell's images. See http://www.pancanal.com/eng/history/murals/index .html, accessed Sept. 22, 2007.

106. *Catalogue of an Exhibition of Lithographs and Etchings of the Panama Canal by Joseph Pennell, September 19 to October 12, 1912* (New York: Frederick Keppel, n.d.), 6–7.

107. Ibid., 5.

108. *Joseph Pennell's Pictures,* 13.

109. The standard biography was written by Pennell's wife, Elizabeth Robins Pennell, *The Life and Letters of Joseph Pennell* (Boston: Little, Brown, 1929).

110. *Joseph Pennell's Pictures,* 7.

111. *Catalogue of an Exhibition,* 5.

112. *Joseph Pennell's Pictures,* 7.

113. *Catalogue of an Exhibition,* 8, and *Joseph Pennell's Pictures,* n.p.

114. J. B. Bishop, *Panama Gateway,* 212.

115. *Joseph Pennell's Pictures,* 3, 7.

116. J. B. Bishop, *Panama Gateway,* 312.

117. Goethals to Pennell, Aug. 7, 1912, cited in *Catalogue of an Exhibition,* n.p.

118. "Mightier than Egypt's tombs, / Fairer than Grecia's, Roma's temples, / Prouder than Milan's statued, spired cathedral, / More picturesque than Rhenish castle-keeps, / We plan even now to raise, beyond them all, / Thy great cathedral sacred industry, no tomb, / A keep for life for practical invention." Walt Whitman, "Song of the Exposition," in *Leaves of Grass: The 1892 Edition* (1892; New York: Bantam Books, 1983), 161.

119. See McCullough, *Path between the Seas,* 604.

120. *Current Literature* 53 (Nov. 1912): 579.

121. See Harvey Green, "Pasteboard Masks: The Stereograph in American Culture, 1865–1910," in Earle, *Points of View,* 113.

122. Allan Sekula's essay "The Body and the Archive" was one of the first explorations of the subject. Numerous studies of the race and gender aspects of visual archives ensued. Recent studies include Shawn Michelle Smith, *American Archives: Gender, Race and Class in Visual Culture* (Princeton, NJ: Princeton University Press, 1999); Elspeth H. Brown, *The Corporate Eye: Photography and the Rationalization of American Commercial Culture, 1884–1929* (Baltimore: Johns Hopkins University Press, 2005); and Anna Pegler Gordon, "In Sight of America: Photography and U.S. Immigration Policy, 1880–1930" (Ph.D. diss., University of Michigan, 2002), which stresses the instability of photography's disciplinary power.

123. Holmes, "Stereoscope and the Stereograph." 81.

124. S. M. Smith, *American Archives,* 4.

125. F. B. Maltby, Principal Assistant Engineer, to Hallen, Feb. 7, 1907; Official Photographs; Policy, Records, Instructions to Photographers, Etc.— General, File 56-A-7, part 1, GC 1904–14, RG 185, NACP.

126. *Canal Record,* Dec. 12, 1908, 115.

127. Hall and Chester, *Panama and the Canal,* 166.

128. Abbot, *Panama and the Canal,* 121.

129. Hall and Chester, *Panama and the Canal,* 75.

130. Abbot, *Panama and the Canal,* 305.

131. An example is a naked boy with folded arms frowning at the camera, shown in "A Primitive Laundry on the Isthmus of Panama," American

Stereoscopic Company stereograph no. 4221719; Stereo File, Foreign Geographical, Panama; Prints and Photographs Division, Library of Congress.

132. Haskin, *Panama Canal,* 274.

133. Abbot, *Panama and the Canal,* 43.

134. Ibid., 313, 315.

135. Anne McClintock, *Imperial Leather: Race, Gender and Sexuality in the Colonial Contest* (New York: Routledge, 1995), 44.

136. Mirzoeff, *Introduction to Visual Culture,* 5.

137. Hugh C. Weir, *With the Flag at Panama: A Story of the Building of the Panama Canal* (Boston: W. A. Wilde, 1911), 6. This second book by Weir is a fictionalized account of the Canal building.

138. Foucault, "Eye of Power," 151.

Chapter 4. Ideal Community

1. John Foster Carr, "The Panama Canal, History—Conditions—Prospects," *Outlook,* April 28, 1906, 963.

2. Hugh C. Weir, *The Conquest of the Isthmus: The Men Who Are Building the Panama Canal—Their Daily Lives, Perils, and Adventures* (New York: G. P. Putnam's Sons, 1909), 9.

3. *Joseph Pennell's Pictures of the Panama Canal: Reproductions of a Series of Lithographs Made by Him on the Isthmus of Panama, January–March, 1912, Together with Impressions and Notes by the Artist* (Philadelphia: J. B. Lippincott, 1912), x.

4. *Canal Record,* Aug. 26, 1908, 412–413.

5. *Canal Record,* July 24, 1912, 384–385.

6. *Canal Record,* Sept. 6, 1911, 12.

7. *Canal Record,* Aug. 26, 1908, 412.

8. *Canal Record,* July 24, 1912, 385.

9. John Abbot, *Panama and the Canal in Picture and Prose: A Complete Story of Panama, as Well as the History, Purpose and Promise of Its World-Famous Canal—the Most Gigantic Engineering Undertaking since the Dawn of Time* (New York: Syndicate, 1913), 321.

10. Ibid., 322.

11. Weir, *Conquest of the Isthmus,* 17.

12. Abbot, *Panama and the Canal,* 323.

13. Weir, *Conquest of the Isthmus,* 28.

14. Arthur Bullard [Albert Edwards], *Panama: The Canal, the Country and the People,* rev. ed. with additional chapters (1911; New York: Macmillan, 1914), 569.

15. Farnham Bishop, *Panama Past and Present* (New York: Century, 1913), 196. This credit system was not unusual for workers' towns run by railroads and mining corporations. It was also used by the Suez Canal Company. See

Zachary Karabell, *Parting the Desert: The Creation of the Suez Canal* (London: John Murray, 2003), 163.

16. F. Bishop, *Panama Past and Present*, 199.

17. For overviews of this development see Susan Strasser, *Satisfaction Guaranteed: The Making of the American Mass Market* (New York: Pantheon Books, 1989); Simon J. Bronner, ed., *Consuming Visions: Accumulation and Display of Goods in America, 1880–1920* (New York: W. W. Norton, 1989); and T. J. Jackson Lears, *Fables of Abundance: A Cultural History of Advertising in America* (New York: Basic Books, 1994).

18. Bullard, *Panama*, 570.

19. Within the United States, similar interpretations of the new culture of consumption were widespread. See, for example, Simon J. Bronner, "Reading Consumer Culture," in Bronner, *Consuming Visions*, 51, and George Cotkin, *Reluctant Modernism: American Thought and Culture, 1880–1900* (New York: Twayne, 1992), xiii, 117.

20. William Rufus Scott, *The Americans in Panama* (New York: Statler, 1912), 190. See also Frederic Jennings Haskin, *The Panama Canal* (Garden City, NY: Doubleday, Page, 1913), 166.

21. See Joseph Bucklin Bishop, *The Panama Gateway* (New York: Charles Scribner's Sons, 1913), 280.

22. Weir, *Conquest of the Isthmus*, 120, and Ralph Emmett Avery, *America's Triumph at Panama: Panorama and Story of the Construction and Operation of the World's Giant Waterway from Ocean to Ocean* (Chicago: Regan Printing House, 1913), 125. A troop of two hundred policemen made sure these orders were followed.

23. See Haskin, *Panama Canal*, 180, and J. B. Bishop, *Panama Gateway*, 280.

24. Ira E. Bennett, *History of the Panama Canal: Its Construction and Builders* (Washington, DC: Historical, 1915), 170.

25. "Tourists and Touristesses: An Interview with Wm. M. Baxter, Official Guide," in *Society of the Chagres, Year Book 1913* (Culebra, CZ: John O. Collins, n.d.), 60. For tourism figures, see p. 70.

26. See *Canal Record*, Dec. 4, 1912, 121.

27. Bennett, *History of the Panama Canal*, 321.

28. For a similar analysis of tourism, see James B. Gilbert, *Perfect Cities: Chicago's Utopias of 1893* (Chicago: University of Chicago Press, 1991), 21.

29. Mary L. McCarty, *Glimpses of Panama and the Canal* (Kansas City: Tiernan-Dart, 1913), 130.

30. See Katherine Emma Manthorne, *Tropical Renaissance: North American Artists Exploring Latin America, 1839–1879* (Washington, DC: Smithsonian Institution Press, 1989), 11, and Stephen Frenkel, "Jungle Stories: American Representations of Tropical Panama," *Geographical Review* 86 (July 1996): 324.

31. Abbot, *Panama and the Canal*, 9.

32. Scott, *Americans in Panama*, 17.

33. Charles Francis Adams, *The Panama Canal Zone: An Epochal Event in Sanitation* (Boston: Massachusetts Historical Society, 1911), 7.

34. Bennett, *History of the Panama Canal,* 170. McCarty, *Glimpses of Panama,* 163, makes an almost identical comment: "It was certainly a good looking and attractive crowd."

35. Bullard, *Panama,* 512. Stephen Frenkel, in his article "Geographical Representations of the 'Other': The Landscape of the Panama Canal Zone," *Journal of Geographical History* 28, no. 1 (2002): 85–99, traces the evolution of a suburban landscape in the Canal Zone. Houses could only be painted in authorized colors to ensure a homogeneous look.

36. Bullard, *Panama,* 527.

37. Avery, *America's Triumph,* 104. See also the description in Porter's Progress of Nations series, cited in John O. Collins, *The Panama Guide* (Panama: Vibert and Dixon, 1912), 48: "In the Canal Zone there are not enough people of any one industrial class, with common church, professional, and cultural interests, to form these little cliques for social stagnation, and the result is a broadening of social and intellectual horizon that keeps most of them in a fever of excitement."

38. Walter Leon Pepperman, *Who Built the Panama Canal?* (London: J. M. Dent and Sons, 1915), 218.

39. See Michael L. Conniff, *Black Labor on a White Canal: Panama, 1904–1981* (Pittsburgh: University of Pittsburgh Press, 1985), 30.

40. Logan Marshall, *The Story of the Panama Canal* (Philadelphia: John C. Winston, 1913), 150.

41. Bennett, *History of the Panama Canal,* 166. Special lodgings for married Caribbean workers did not exist.

42. A. W. Wyndham, *The Panama Canal* (New York: Howard F. Curtis, 1907), 27.

43. Haskin, *Panama Canal,* 154.

44. Harry A. Franck, *Zone Policeman 88: A Close Range Study of the Panama Canal and Its Workers* (New York: Century, 1913), 43. He chided the Canal Zone as a place "with outward democracy and inward caste" (221).

45. F. Bishop, *Panama Past and Present,* 203.

46. Alfred B. Hall and Clarence L. Chester, *Panama and the Canal* (New York: Newson, 1910), 158.

47. Gail Bederman, *Manliness and Civilization: A Cultural History of Gender and Race in the United States, 1880–1917* (Chicago: University of Chicago Press, 1995), 28.

48. Abbot, *Panama and the Canal,* 19. For a similar assessment during the early construction period, see John Foster Carr, "The Panama Canal: The Silver Men," *Outlook,* May 19, 1906, 118: "Tropical nature has no penalties for the improvident, and even the bee, they say, stores no honey here. Little or no work

brings all the necessities of life, and in his good-humored ignorance the West Indian is generally well content to do without the rest."

49. Weir, *Conquest of the Isthmus*, 78.

50. Frenkel, "Jungle Stories," 321, states that positive views of Panama were often associated with travel and negative impressions with living in the tropics.

51. Bullard, *Panama*, 10.

52. Michael Adas, *Dominance by Design: Technological Imperatives and America's Civilizing Mission* (Cambridge, MA: Belknap Press of Harvard University Press, 2006), 155–156.

53. J. B. Bishop, *Panama Gateway*, 307.

54. Weir, *Conquest of the Isthmus*, 178. On the women in the Canal Zone, see Paul W. Morgan Jr., "The Role of North American Women in U.S. Cultural Chauvinism in the Panama Canal Zone, 1904–1945" (Ph.D. diss., Florida State University, 2000).

55. Haskin, *Panama Canal*, 180.

56. Bullard, *Panama*, 538. He made an almost identical comment in an article for a popular women's magazine. See Arthur Bullard, "The Romance of the Canal," *Ladies' Home Journal* 31 (Oct. 1914): 93.

57. See Alexander Schmidt, *Reisen in die Moderne: Der Amerika-Diskurs des deutschen Bürgertums vor dem Ersten Weltkrieg im europäischen Vergleich* (Berlin: Akademie Verlag, 1997), esp. chap. 1.

58. Bullard, *Panama*, 567.

59. Ibid., 558.

60. Ibid., 508.

61. "Work and Welfare on the Canal," *Independent*, April 9, 1909, 914.

62. Abbot, *Panama and the Canal*, 329.

63. "Work and Welfare on the Canal," 913.

64. Franck, *Zone Policeman 88*, 204.

65. Bullard, *Panama*, 204.

66. J. B. Bishop, *Panama Gateway*, 258.

67. Abbot, *Panama and the Canal*, 326–327, cited a businessman, "a very distinguished financier of New York"; Bullard, *Panama*, 577, cited a Canal worker who was a member of the Socialist Party of America.

68. Henry Kitchell Webster, "Real Socialism," *Atlantic Monthly* 111 (May 1913): 634.

69. Ibid., 636.

70. Roosevelt to Abbott, June 6, 1908, in *The Letters of Theodore Roosevelt*, ed. Elting E. Morison, vol. 6 (Cambridge, MA: Harvard University Press, 1951), 1081.

71. Ibid.

72. Scott, *Americans in Panama*, 152 and 153.

73. Bullard, *Panama*, 572.

74. In-text citations refer to Edward Bellamy, *Looking Backward: 2000–1887* (1888; New York: Penguin Books, 1960). Over the past decades, a sizeable number of studies on Bellamy and other utopian writers have been published. For classic biographies, see Arthur E. Morgan, *Edward Bellamy* (New York: Columbia University Press, 1944); and Sylvia E. Bowman, *Edward Bellamy* (Boston: Twayne, 1986). For important scholarship on Bellamy's influence, see Sylvia E. Bowman, ed., *Edward Bellamy Abroad: An American Prophet's Influence* (New York: Twayne, 1962); Arthur Lipow, *Authoritarian Socialism in America: Edward Bellamy and the Nationalist Movement* (Berkeley: University of California Press, 1982); Daphne Patai, ed., *Looking Backward, 1988–1888: Essays on Edward Bellamy* (Amherst: University of Massachusetts Press, 1988); and, most recently, James B. Gilbert, "Social Utopias in Modern America," in *Visions of the Future in Germany and America*, ed. Norbert Finzsch and Hermann Wellenreuther (Oxford: Berg, 2001), 251–274. On the broader context of utopian writers, communities, and political movements before and after Bellamy see Kenneth M. Roemer, *The Obsolete Necessity: America in Utopian Writings, 1888–1900* (Kent, OH: Kent State University Press, 1976); John F. Kasson, *Civilizing the Machine: Technology and Republican Values in America, 1776–1900* (New York: Grossman, 1976); John L. Thomas, *Alternative America: Henry George, Edward Bellamy, Henry Demarest Lloyd, and the Adversary Tradition* (Cambridge, MA: Belknap Press of Harvard University Press, 1983); Rob Kroes, ed., *Nineteen Eighty-Four and the Apocalyptic Imagination in America* (Amsterdam: Free University Press, 1985); Howard P. Segal, *Technological Utopianism in American Culture* (Chicago: University of Chicago Press, 1985); and Edward K. Spann, *Brotherly Tomorrows: Movements for a Cooperative Society in America, 1820–1920* (New York: Columbia University Press, 1989).

75. Abbot, *Panama and the Canal*, 330.

76. Franck, *Zone Policeman 88*, 216.

77. Herbert Knapp and Mary Knapp, *Red, White, and Blue Paradise: The American Canal Zone in Panama* (San Diego: Harcourt Brace Jovanovich, 1984), 4. The two teachers call the Canal Zone "the world's first workers' paradise—a star-spangled red, white, and blue postcapitalist society that was established while Russia was still groaning under the yoke of the Czar" (3).

78. The radio effect was achieved by simultaneously playing orchestras. Through telephone lines, their live music was broadcast into the homes. The radio could even be programmed as an alarm clock. See Bellamy, *Looking Backward*, 87, 103.

79. James B. Gilbert, "Inventing the Twentieth Century," unpublished essay, n.p. The manuscript was provided by the author.

80. Bellamy, *Looking Backward*, 212–213. His enthusiasm was rooted in a personal failure: the elite military academy West Point had turned down his application due to poor health.

81. J. B. Bishop, *Panama Gateway*, 195. For the full quote, see p. 108.

82. Roemer, *Obsolete Necessity*, 2.

83. Spann, *Brotherly Tomorrows*, 190.

84. Segal, *Technological Utopianism*, 102.

85. Susan Matarese, *American Foreign Policy and the Utopian Imagination* (Amherst: University of Massachusetts Press, 2001), 14.

86. "Come Muse migrate from Greece and Ionia / . . . For know a better, fresher, busier sphere, a wide, untried domain awaits, demands you." Walt Whitman, *Leaves of Grass: The 1892 Edition* (1892; New York: Bantam Books, 1983), 159. The poem, written in 1871, had been commissioned for an industrial exposition in New York.

87. See Lipow, *Authoritarian Socialism in America*, 39. Bellamy was eighteen when he traveled through Europe in 1868.

88. Bellamy, *Looking Backward*, 100.

89. See Spann, *Brotherly Tomorrows*, 190. In Germany, the first translation bore the notable subtitle *Alles Verstaatlicht* (Everything Nationalized). Numerous versions were published in 1890, one of them prepared by the socialist reformer Clara Zetkin. See Franz X. Riederer, "The German Acceptance and Reaction," in Bowman, *Edward Bellamy Abroad*, 152–155.

90. Lipow, *Authoritarian Socialism in America*, 30, estimates the figure at 200,000, while Cotkin, *Reluctant Modernism*, 110, mentions 325,000. Spann, *Brotherly Tomorrows*, 190, puts the total number of copies sold within the first ten years at 400,000.

91. See Lipow, *Authoritarian Socialism in America*, 30–33, and Wiebe, *Search for Order*, 71–75. The Populists sprung from an agrarian movement and were organized in the People's Party from 1892 to 1896. The Social Democracy of America was founded in 1897 and was later renamed Social Democratic Party of America. In 1901 it merged with parts of the Socialist Labor Party to form the Socialist Party of America.

92. See Spann, *Brotherly Tomorrows*, 203. Bellamy died in 1898 at the age of forty-eight.

93. Cited in Cecilia Elizabeth O'Leary, *To Die For: The Paradox of American Patriotism* (Princeton, NJ: Princeton University Press, 1999), 160.

94. Bellamy to Howells, June 17, 1888, cited in Lipow, *Authoritarian Socialism in America*, 22.

95. "No historical authority nowadays doubts that they were paid by the great monopolies to wave the red flag and talk about burning, sacking, and blowing people up, in order, by alarming the timid, to head off any real reforms," recounts Dr. Leete. Bellamy, *Looking Backward*, 170–171.

96. Cotkin, *Reluctant Modernism*, 111.

97. See Lipow, *Authoritarian Socialism in America*, 36, and Segal, *Technological Utopianism*, 30.

98. Roemer, *Obsolete Necessity*, 138. See also Gilbert, "Inventing the Twentieth Century," n.p.

99. Haskin, *Panama Canal*, 193.

100. Gilbert, "Inventing the Twentieth Century," n.p.

101. Abbot, *Panama and the Canal*, 326.

102. Bullard, *Panama*, 578.

103. See Dirk van Laak, *Weiße Elefanten: Anspruch und Scheitern technischer Großprojekte im 20. Jahrhundert* (Stuttgart: Deutsche Verlags-Anstalt, 1999), 32–33, and Matt K. Matsuda, *Empire of Love: Histories of France and the Pacific* (Oxford: Oxford University Press, 2005), 51–53.

104. See Gilbert, "Social Utopias," 256, 264, and A. Morgan, *Edward Bellamy*, xii. "Various New Deal policies and proposals would seem to have been taken almost directly from the pages of *Looking Backward*," Morgan notes. The 1930s also saw an attempt at building utopian communities on a wider scale. The suburban Greenbelt towns "promoted a sort of pluralist homogeneity" through direct and equal participation while being open only to middle-class families. See Gilbert, "Social Utopias," 254–255, 266–267. On Soviet public projects, see Klaus Gestwa, "Technik als Kultur der Zukunft: Der Kult um die Stalinschen Großbauten des Kommunismus," *Geschichte und Gesellschaft* 30, no. 1 (2004): 37–73.

105. See Thomas P. Hughes, *American Genesis: A Century of Invention and Technological Enthusiasm, 1870–1970* (New York: Penguin Books, 1989), 367. The general objective was to transform a poverty-stricken valley into a region for economic growth through flood control and hydroelectrical power.

106. John Major, *Prize Possession: The United States and the Panama Canal, 1903–1979* (Cambridge, UK: Cambridge University Press, 1993), 70.

107. Abbot, *Panama and the Canal*, 329. For use of the term, see also J. B. Bishop, *Panama Gateway*, 259.

108. Avery, *America's Triumph*, n.p. The quote is taken from a photo caption in the front of the book.

109. J. B. Bishop, *Panama Gateway*, 283.

110. Ibid.

111. Farnham Bishop, "The End of the Big Job," *Century Magazine* 85 (Dec. 1912): 274.

112. Marshall, *Story of the Panama Canal*, 135.

113. Franck, *Zone Policeman 88*, 197.

114. Ibid., 205.

115. On Goethals's background, see Ray Stannard Baker, "Goethals: The Man and How He Works," *American Magazine* 76 (Oct. 1913): 22–27, and Joseph Bucklin Bishop and Farnham Bishop, *Goethals, Genius of the Panama Canal: A Biography* (New York: Harper and Brothers, 1930), published shortly after

Goethals's death in 1928. A recent study by Walt Griffin, "George W. Goethals and the Panama Canal" (Ph.D. diss., University of Cincinnati, 1988), adds only a few critical insights to the Bishops' official work.

116. See George W. Goethals, "The Building of the Panama Canal: Organization of the Force," *Scribner's Magazine* 57 (May 1915): 538.

117. See Major, *Prize Possession*, 71–72.

118. F. Bishop, *Panama Past and Present*, 274.

119. McCarty, *Glimpses of Panama*, 121.

120. See Joseph Bucklin Bishop, "The Personality of Colonel Goethals," *Scribner's Magazine* 57 (Feb. 1915): 134, 145.

121. She illustrates the tense relationship between the two men with an anecdote: "'Do you know, Gorgas,' Colonel Goethals said one day, 'that every mosquito you kill costs the United States Government ten dollars?' — 'But just think,' answered Gorgas, 'one of those ten-dollar mosquitos might bite you, and what a loss that would be to the country!'" Marie D. Gorgas and Burton J. Hendrick, *William Crawford Gorgas: His Life and Work* (Garden City, NY: Doubleday, Page, 1924), 222. On Goethals's perspective of the conflict, see Goethals, "Building of the Panama Canal," 545.

122. Bellamy, *Looking Backward*, 165.

123. Theodore Roosevelt, *An Autobiography* (1913; New York: Da Capo Press, 1985), 543.

124. Bullard, *Panama*, 574.

125. Frank Morton Todd, *The Story of the Exposition: Being the Official History of the International Celebration Held at San Francisco in 1915 to Commemorate the Discovery of the Pacific Ocean and the Construction of the Panama Canal* vol. 3 (New York: G. P. Putnam's Sons, 1921), 135.

126. Goethals, "Building of the Panama Canal," 531.

127. J. B. Bishop, "Personality of Colonel Goethals," 149.

128. See ibid., 131, and F. Bishop, "End of the Big Job," 274.

129. F. Bishop, *Panama Past and Present*, 225–226.

130. John Foster Fraser, *Panama and What It Means* (London: Cassell, 1913), 84, quotes the lyrics: "If you have any cause to kick, or feel disposed to howl, / If things ain't running just to suit, and there's a chance to growl, / If you have any axe to grind or graft to shuffle through, / Just put it up to Colonel G. like all the others do. / See Colonel Goethals, tell Colonel Goethals, / It's the only right and proper thing to do. / Just write a letter, or, even better, / Arrange a little Sunday interview."

131. Peter C. MacFarlane, "The Solomon of the Isthmus," *Collier's*, Dec. 12, 1911, 27–28.

132. See F. Bishop, *Panama Past and Present*, 225. Under Harun's reign (ca. 763–809), the city of Baghdad reached the height of its political and cultural power.

133. See M. Gorgas and Hendrick, *William Crawford Gorgas*, 219.

134. Goethals, "Building of the Panama Canal," 726.

135. A photograph of the three-hundred-pound governor and his family in matching attire is reproduced in R. Jackson Wilson, James Gilbert et al., *The Pursuit of Liberty: A History of the American People*, vol. 2 (New York: HarperCollins, 1996), 261.

136. The colonial project in the Philippines has received renewed scholarly attention in the past decades. See Julian Go and Anne L. Foster, eds., *The American Colonial State in the Philippines: Global Perspectives* (Durham, NC: Duke University Press, 2003); Frank Schumacher, "The American Way of Empire: National Tradition and Transatlantic Adaptation in America's Search for Imperial Identity, 1898–1910," *Bulletin of the German Historical Institute* 31 (Fall 2002); H. W. Brands, *Bound to Empire: The United States and the Philippines* (Oxford: Oxford University Press, 1992); and Glenn Anthony May, *Social Engineering in the Philippines: The Aims, Execution, and Impact of American Colonial Policy, 1900–1913* (Westport, CT: Greenwood Press, 1980). The island group finally gained independence in 1946.

137. G. A. May, *Social Engineering*, 17.

138. See ibid., xvii.

139. Adas, *Dominance by Design*, 136, 144.

140. May, *Social Engineering*, 21.

141. On the role of engineers, see David F. Noble, *America by Design: Science, Technology, and the Rise of Corporate Capitalism* (Oxford: Oxford University Press, 1977), 33–65; Stefan Willeke, *Die Technokratiebewegung in Nordamerika und Deutschland zwischen den Weltkriegen: Eine vergleichende Analyse* (Frankfurt/Main: Peter Lang, 1995), 31–55; and Ruth Oldenziel, *Making Technology Masculine: Men, Women and Modern Machines in America, 1870–1945* (Amsterdam: Amsterdam University Press, 1999).

142. Oldenziel, *Making Technology Masculine*, 119.

143. Ray Stannard Baker, "Frederick W. Taylor—Scientist in Business Management," *American Magazine* 71 (March 1911): 570.

144. See Noble, *America by Design*, 39.

145. See Hughes, *American Genesis*, 188–203, 250–259; and Willeke, *Die Technokratiebewegung*, 42. On Taylor, see also Robert Kanigel, *The One Best Way: Frederick Winslow Taylor and the Enigma of Efficiency* (New York: Viking, 1997).

146. Cited in Willeke, *Die Technokratiebewegung*, 55. The anti-democratic New Machine association was founded in 1917 by the engineer and Taylor-adherent Henry L. Gantt in order to criticize and radicalize President Wilson's mobilization policies. It is often seen as a predecessor to the technocracy movement of the 1920s and 1930s.

147. See Percy MacKaye, "Goethals," *Literary Digest* 50, Jan. 23, 1915, 155.

148. Todd, *Story of the Exposition*, 5:85.

149. See Scott, *Americans in Panama*, 193. In the *Canal Record*, Nov. 23, 1910, 97, it is reported that the boilermakers handed in resignations.

150. See Marshall, *Story of the Panama Canal*, 154, and Major, *Prize Possession*, 80.

151. Goethals, "Building of the Panama Canal," 413.

152. See J. B. Bishop, "Personality of Colonel Goethals," 148–149, and Baker, "Goethals," 23.

153. Scott, *Americans in Panama*, 194.

154. J. B. Bishop, *Panama Gateway*, 194.

155. Porter's Progress of Nations series, cited in Collins, *Panama Guide*, 49.

156. Cited in David E. Nye, *American Technological Sublime* (Cambridge, MA: MIT Press, 1994), 36.

157. Ray Stannard Baker, "The Glory of Panama: How the Big Ditch, Dug on Honor, Is a Great Example of the New Idealism in Public Service," *American Magazine* 76 (Nov. 1913): 33–37. Additional quotations from this article are cited parenthetically in the text.

158. See Cecilia Tichi, *Exposés and Excess: Muckraking in America, 1900/2000* (Philadelphia: University of Pennsylvania Press, 2004), 3.

159. On Bullard, see Peter Conn, *The Divided Mind: Ideology and Imagination in America, 1898–1917* (Cambridge, UK: Cambridge University Press, 1983), 98–103, and Mark Pittenger, *American Socialists and Evolutionary Thought, 1870–1920* (Madison: University of Wisconsin Press, 1993).

160. On the public intellectuals, see Daniel Rodgers, *Atlantic Crossings: Social Politics in a Progressive Age* (Cambridge, MA: Belknap Press of Harvard University Press, 1998); Brian Lloyd, *Left Out: Pragmatism, Exceptionalism, and the Power of American Marxism, 1890–1922* (Baltimore: Johns Hopkins University Press, 1997); Christopher Lasch, *The True and Only Heaven: Progress and Its Critics* (New York: W. W. Norton, 1991); Casey Blake, *Beloved Community: The Cultural Criticism of Randolph Bourne, Van Wyck Brooks, Waldo Frank, and Lewis Mumford* (Chapel Hill: University of North Carolina Press, 1990); Walter Benn Michaels, "An American Tragedy, or the Promise of American Life," *Representations* 5 (Winter 1989): 71–98; Mark Pittenger, "Science, Culture and the New Socialist Intellectuals before World War I," *American Studies* 27 (Spring 1987): 73–91; James Kloppenburg, *Uncertain Victory: Social Democracy and Progressivism in European and American Thought, 1870–1920* (New York: Oxford University Press, 1986), and James Gilbert, *Designing the Industrial State: The Intellectual Pursuit of Collectivism in America, 1880–1940* (Chicago: Quadrangle Books, 1972). Charles Forcey, *The Crossroads of Liberalism: Croly, Weyl, Lippmann, and the Progressive Era, 1900–1925* (Oxford: Oxford University Press, 1961), remains important for its wealth of details.

161. Walter Lippmann, *Drift and Mastery: An Attempt to Diagnose the Current Unrest* (1914; Madison: University of Wisconsin Press, 1985), 111.

162. Baker, "Glory of Panama," 37.

163. See Pittenger, "Science, Culture and the New Socialist Intellectuals," 77.

164. Gilbert, "Designing the Industrial State," 23.

165. Walter Edward Weyl, *The New Democracy: An Essay on Certain Political and Economic Tendencies in the United States* (1912; New York: Harper and Row, 1964), 276.

166. Lippmann, *Drift and Mastery*, 174.

167. Bellamy, *Looking Backward*, 56, 54.

168. See Charles P. Steinmetz, *America and the New Epoch* (New York: Harper and Brothers, 1916).

169. William English Walling and Harry W. Laidler, eds., *State Socialism Pro and Con: Official Documents and Other Authoritative Selections—Showing the World-Wide Replacement of Private by Governmental Industry before and during the War* (New York: Henry Holt, 1917), 251. A "State Socialist government in Panama" had been established, the authors argued.

170. Weyl, *New Democracy*, 276.

171. Lippmann, *Drift and Mastery*, 50.

172. Ibid., 171.

173. William English Walling, "State Socialism and the Individual," *New Review* 1 (June 1913): 581.

174. On Walling's changing views, see Gilbert, *Designing the Industrial State*, 201, and Lloyd, *Left Out*, 213. In *State Socialism Pro and Con*, Walling suggested that State Socialism "suppresses anti-social individualism most thoroughly" (xliv). Marxists considered it an interim phase between capitalism and true socialism.

175. On Lippmann, see Ronald Steel, *Walter Lippmann and the American Century* (Boston: Little, Brown, 1970).

176. See Forcey, *Crossroads of Liberalism*, 121. On Croly and the *New Republic*, see also 31–39, 169–177.

177. See Ronald Schaffer, *America in the Great War: The Rise of the War Welfare State* (New York: Oxford University Press, 1991), 120–123. The woman eventually committed suicide.

178. On the impact of the war on American society, see David M. Kennedy, *Over Here: The First World War and American Society* (New York: Oxford University Press, 1980); Neil A. Wynn, *From Progressivism to Prosperity: World War I and American Society* (New York: London: Holmes and Meier, 1986) and Schaffer, *America in the Great War*.

179. "Force and Ideas," *New Republic*, Nov. 7, 1914, 7.

180. "The Landslide into Collectivism," *New Republic*, April 10, 1915, 250.

181. Herbert Croly, *The Promise of American Life: With a New Introduction by Scott R. Bowman* (1909; New Brunswick, NJ: Transaction, 1993), 168.

182. See Schaffer, *America in the Great War*, 114; and Kloppenburg, *Uncertain Victory*, 364.

183. Louis C. Fraina, "The Socialism of the Sword," *New Review*, May 15, 1915, 26.

184. Kennedy, *Over Here*, 98. See also Rodgers, *Atlantic Crossings*, 278–279.

185. See Steel, *Walter Lippmann*, 116; and Schaffer, *America in the Great War*, 115. On the Inquiry, see also Neil Smith, *American Empire: Roosevelt's Geographer and the Prelude to Globalization* (Berkeley: University of California Press, 2003), 113–138.

186. Randolph Bourne, "A War Diary," in *War and the Intellectuals*, ed. Carl Resek (New York: Harper Torchbooks, 1964), cited in Steel, *Walter Lippmann*, 114. Bourne is often seen as an unwavering voice who called for a cultural aura of inclusion and internationalism and who did not falter under the temptations of power. See Blake, *Beloved Community*, 78, 116, 134, as well as Randolph Bourne, "Trans-National America," *Atlantic Monthly* 118 (July 1916): 86–97. Bourne died at age thirty-two in 1918.

187. See Kennedy, *Over Here*, 128–135; and Wynn, *From Progressivism to Prosperity*, 69, 77–78. The frame of the WIB was made possible by the Overman Act of 1917, which gave the president the power to reorganize government agencies without authorization from Congress. Nevertheless, the WIB had little formal authority, prompting Kennedy to call it "largely a cosmetic creation, even a political ploy" (128).

188. Wynn, *From Progressivism to Prosperity*, 78.

189. See Kennedy, *Over Here*, 253–258; and Schaffer, *America in the Great War*, 37–39.

190. See Kennedy, *Over Here*, 141; Wynn, *From Progressivism to Prosperity*, xix; and Ellis W. Hawley, *The Great War and the Search for a Modern Order: A History of the American People and Their Institutions, 1917–1933*, 2nd ed. (New York: St. Martin's Press, 1992), 19.

191. Kennedy, *Over Here*, 128.

192. See Schaffer, *America in the Great War*, 5, 15.

193. See ibid., 116.

194. Cited in Forcey, *Crossroads of Liberalism*, 290. See also Schaffer, *America in the Great War*, 117, and Kennedy, *Over Here*, 91–92.

195. See Lasch, *True and Only Heaven*, 364, and Rodgers, *Atlantic Crossings*, 316–317.

196. Walter Lippmann, *Public Opinion* (1922; Mineola, NY: Dover, 2004), 17.

197. Ibid., 15.

198. Ibid., 17.

199. See James M. Skinner, *France and Panama: The Unknown Years, 1894–1908* (New York: P. Lang, 1989), 277.

200. Willis John Abbot, *Watching the World Go By* (Boston: Little, Brown, 1933), 342. Knapp and Knapp, *Red, White, and Blue Paradise*, 77, mention the interview. On a comparison between fascism and the New Deal and their

respective roots, see Wolfgang Schivelbusch, *Three New Deals: Reflections on Roosevelt's America, Mussolini's Italy, and Hitler's Germany, 1933–1939* (New York: Metropolitan Books, 2006). Like Rodgers's *Atlantic Crossings*, the study is a reminder of the transatlantic nature of collectivist impulses, even though fundamental differences remained. What separated all of the Western regimes of the 1930s from the pre–World War I debates was, of course, the Soviet Union as a new counterpart.

201. Baker, "Goethals," 25.

202. "Panama and Socialism," *Nation*, Nov. 13, 1913, 454.

203. Raymond Williams, *Keywords: A Vocabulary of Culture and Society* (Glasgow: Fontana, 1976), 49.

204. Lippmann, *Drift and Mastery*, 170.

205. Edward Bellamy, "Letter to the People's Party," *New Nation*, Oct. 22, 1892, 645. The quote was part of a speech to be delivered at the founding convention of the *People's Party*. Due to illness, Bellamy could not attend and instead published the text as a letter in the official magazine of the *Nationalist Clubs*.

Chapter 5. Celebrating the Canal

1. Farnham Bishop, *Panama Past and Present* (New York: Century, 1913), 230.

2. On the interpretations of world's fairs' admission figures, see James Gilbert's essay "Fair Itineraries: Experience, Memory and the History of the St. Louis World's Fair of 1904," which will be part of a forthcoming book. I thank the author for supplying the manuscript.

3. Exceptions are Bill Brown, "Science Fiction, the World's Fair, and the Prosthetics of Empire, 1910–1915," in *Cultures of United States Imperialism*, ed. Amy Kaplan and Donald E. Pease (Durham, NC: Duke University Press, 1993); Alfred Charles Richard Jr., *The Panama Canal in American National Consciousness, 1870–1990* (New York: Garland, 1990), 233–248; and Alan Dawley, *Changing the World: American Progressives in War and Revolution* (Princeton, NJ: Princeton University Press, 2003), 83–92.

4. See Ralph Emmett Avery, *America's Triumph at Panama: Panorama and Story of the Construction and Operation of the World's Giant Waterway from Ocean to Ocean* (Chicago: Regan, 1913); Ira E. Bennett, *History of the Panama Canal: Its Construction and Builders* (Washington, DC: Historical, 1915); and Frederic Jennings Haskin, *The Panama Canal* (Garden City, NY: Doubleday, Page, 1913). An enlarged edition of Avery's book also included the fair in its title. See Ralph Emmett Avery, *The Panama Canal and Golden Gate Exposition: Authentic and Complete Story of the Building and Operation of the Great Waterway—the Eighth Wonder of the World* (New York: Leslie-Judge, 1915).

5. Robert W. Rydell, John E. Findling, and Kimberly D. Pelle, *Fair America: World's Fairs in the United States* (Washington, DC: Smithsonian Institution Press, 2000), 9.

6. Rydell's *All the World's a Fair* was the path-breaking work and devoted special emphasis to the anthropological exhibits. Over the course of the years and with the help of co-authors, Rydell expanded the scope of his exposition scholarship in terms of methodology, themes, and geography. See Robert W. Rydell, *World of Fairs: The Century-of-Progress Expositions* (Chicago: University of Chicago Press, 1993); Robert W. Rydell and Nancy Gwinn, eds., *Fair Representations: World's Fairs and the Modern World* (Amsterdam: VU University Press, 1994); and, most recently, Robert W. Rydell, John E. Findling, and Kimberly D. Pelle, *Fair America: World's Fairs in the United States* (Washington, DC : Smithsonian Institution Press, 2000). Important general studies and reference works by other authors include Paul Greenhalgh, *Ephemeral Vistas: The Expositions Universelles, Great Exhibitions and World's Fairs, 1851–1939* (Manchester: Manchester University Press, 1988); John E. Findling, ed., *Historical Dictionary of World's Fairs and Expositions, 1851–1988* (Westport, CT: Greenwood Press, 1990), Winfried Kretschmer, *Geschichte der Weltausstellungen* (Frankfurt/Main: Campus Verlag, 1999); and Paul A. Tencotte, "Kaleidoscopes of the World: International Exhibitions and the Concept of Culture-Place, 1851–1915," *American Studies* 28 (Spring 1987): 5–30. For bibliographical material on the individual fairs, see Robert W. Rydell, "The Literature of the International Expositions," in *The Books of the Fairs: Material about World's Fairs, 1834–1916, in the Smithsonian Institution Libraries* (Chicago: American Library Association, 1992), 1–64, as well as the comprehensive on-line resource by Alexander C. T. Geppert, Jean Coffey, and Tammy Lau, "International Exhibitions, Expositions Universelles and World's Fairs, 1851–1951: A Bibliography," *Wolkenkuckucksheim: Internationale Zeitschrift für Theorie und Wissenschaft* (special issue, 2000), http://www.theo.tu-cottbus.de/Wolke/eng/Bibliography/ExpoBibliography.htm, accessed Sept. 22, 2007. The best-studied exposition both inside and outside the United States remains the Columbian Exposition, for which more than 150 entries are listed.

7. Rydell, *All the World's a Fair*, 295.

8. Rydell, *Fair Representations*, 22.

9. For the complex functions of early amusement parks, see John F. Kasson, *Amusing the Million: Coney Island at the Turn of the Century* (New York: Hill and Wang, 1978).

10. See Burton Benedict, "The Anthropology of World's Fairs," in *The Anthropology of World's Fairs: San Francisco's Panama-Pacific International Exposition of 1915*, ed. Burton Benedict (Berkeley, CA: Scolar Press, 1983), 3.

11. Benedict's edited volume *The Anthropology of World's Fairs* was published even before Rydells's *All the World's a Fair* but remains the only book-length academic study of the PPIE. More specific are Abigail Margaret Markwyn, "Constructing 'an Epitome of Civilization': Local Politics and Visions of Progressive Era America at San Francisco's Panama-Pacific International Exposition" (Ph.D. diss., University of Wisconsin, 2006); and Portia Lee, "Victorious Spirit:

Regional Influences in the Architecture, Landscaping and Murals of the Panama-Pacific International Exposition" (Ph.D. diss., George Washington University, 1984). Donna Ewald and Peter Clute, *San Francisco Invites the World: The Panama-Pacific International Exposition of 1915* (San Francisco: Chronicle Books, 1991), and Wayne Bonnett, *City of Dreams: Panama-Pacific Exposition 1915* (Sausalito, CA: Windgate Press, 1995), are popular books notable for their reproductions of historic photographs and illustrations. The University of California organized an exhibition on fairs (among them the PPIE) and other public events in California history, which was accompanied by the noteworthy booklet *Looking Backward, Looking Forward: Visions of the Golden State, The Bancroft Library, University of California, Berkeley, March 17–September 4, 2000* (n.p., n.d.). Contemporary sources and exhibits are listed in Burton Benedict, M. Miriam Dobkin, and Elizabeth Armstrong, *A Catalogue of Posters, Photographs, Paintings, Drawings, Furniture, Documents, Souvenirs, Statues, Books, Medals, Dolls, Music Sheets, Postcards, Curiosities, Banners, Awards, Memoirs, Etcetera from San Francisco's Panama Pacific International Exposition 1915, October 30, 1982–December 31, 1983, the Lowie Museum of Anthropology, University of California, Berkeley* (Berkeley: University of California Press, 1982), as well as in Marvin R. Nathan, *San Francisco's International Expositions: A Bibliography, Including Listings for the Mechanics' Institute Exhibitions* (San Francisco: San Francisco State University, 1990). The most comprehensive primary source for the PPIE is Frank Morton Todd's five-volume history, *The Story of the Exposition: Being the Official History of the International Celebration Held at San Francisco in 1915 to Commemorate the Discovery of the Pacific Ocean and the Construction of the Panama Canal* (New York: G. P. Putnam's Sons, 1921), which was published after the war and therefore contains some interpretations made in hindsight. Another massive volume, which has been completely ignored in studies of the fair, is the *History of the Panama-Pacific International Exposition, Comprising the History of the Panama Canal and a Full Account of the World's Greatest Exposition, Embracing the Participation of the States and Nations of the World and other Events at San Francisco, 1915, Compiled by the Pan-Pacific Press Association, Ltd.* (n.d.). Most likely, it came out before Todd's official history. The editors used Ernest Hallen's photographs for their illustrations of the Panama Canal.

    12. Rydell, *Fair Representations*, 1.

    13. Rydell, "Literature of the International Expositions," 2.

    14. See Benedict, "Anthropology of World's Fairs," 67. On Californian and San Francisco history and their impact on the PPIE, see Kevin Starr, *Americans and the California Dream, 1850–1915* (Oxford: Oxford University Press, 1973), and his subsequent *Inventing the Dream: California through the Progressive Era* (Oxford: Oxford University Press, 1985); Judd Kahn, *Imperial San Francisco: Politics and Planning in an American City, 1897–1906* (Lincoln: University of Nebraska Press, 1979); William Issel and Robert W. Cherny, *San Francisco, 1865–1932: Politics, Power, and Urban Development* (Berkeley: University of California Press, 1986);

Gray Brechin, *Imperial San Francisco: Urban Power, Earthly Ruin* (Berkeley: University of California Press, 1999), an environmental history, and also Brechin's essay "San Francisco: The City Beautiful," in *Visionary San Francisco*, ed. Paolo Polledri (San Francisco: San Francisco Museum of Modern Art, 1990), 40–62.

15. Cited in Todd, *Story of the Exposition*, 1:35.

16. See Starr, *Americans and the California Dream*, 288–290; Brechin, "San Francisco," 49–50; Kahn, *Imperial San Francisco*, 2; and *Looking Backward, Looking Forward*.

17. See Marie Bolton, "Recovery for Whom? Social Conflict after the San Francisco Earthquake and Fire, 1906–1915" (Ph.D. diss., University of California, Davis, 1997), 1.

18. See Starr, *Americans and the California Dream*, 294.

19. On this exposition, the first world's fair held on the West Coast, see Arthur Chandler and Marvin Nathan, *The Fantastic Fair: The Story of the California Midwinter International Exposition, Golden Gate Park, San Francisco, 1894* (St. Paul, MN: Pogo Press, 1993).

20. Issel and Cherny, *San Francisco*, 168.

21. Haskin, *Panama Canal*, 376. The best overview of the PCE, which was also held in 1915, is Richard W. Amero, "The Making of the Panama-California Exposition, 1909–1915," *Journal of San Diego History* 36, no. 1 (1990): 1–47. The Panama authors usually focused on the PPIE and sometimes devoted a shorter chapter to the exposition in San Diego.

22. For the following, see Issel and Cherny, *San Francisco*, 161–165, and Bolton, "Recovery for Whom?"

23. See Brechin, "San Francisco," 54–55.

24. See Hamilton Wright, "Opening the Panama-Pacific Exposition," *Overland Monthly* 65 (March 1915): 197–198.

25. See *Official Guide of the Panama-Pacific International Exposition 1915*, first ed. (San Francisco: Wahlgreen, 1915), 15.

26. See *Looking Backward, Looking Forward*.

27. See G. Allen Greb, "Opening a New Frontier: San Francisco, Los Angeles and the Panama Canal, 1900–1914," *Pacific Historical Quarterly* 47, no. 3 (1978): 411.

28. Cited in Bennett, *History of the Panama Canal*, 303.

29. Marjorie M. Dobkin, "A Twenty-Five-Million-Dollar Mirage," in Benedict, *Anthropology of World's Fairs*, 70. See also Starr, *Americans and the California Dream*, 296.

30. Bolton, "Recovery for Whom?" 3, 190–192.

31. Benjamin Ide Wheeler, "The Meaning of the Canal," *American Review of Reviews* 51 (Feb. 1915): 164. Wheeler was president of the University of California and a friend of Roosevelt's.

32. See Benedict, Dobkin, and Armstrong, *A Catalogue of Posters*.

33. See Rydell, *All the World's a Fair*, 69.

34. Marshall, *Seeing America*, 238.

35. Jesse Lynch Williams, "The Color Scheme at the Panama-Pacific International Exposition," *Scribner's Magazine* 56 (Sept. 1914): 277.

36. Ben Macomber, *The Jewel City: Its Planning and Achievement; Its Architecture, Sculpture, Symbolism, and Music; Its Gardens, Palaces, and Exhibits* (San Francisco: John H. Williams, 1915), 45.

37. Wright, "Opening the Panama-Pacific Exposition," 198. For a discussion of the light effects at the fair, see David E. Nye, *Electrifying America: Social Meanings of a New Technology, 1880–1940* (Cambridge, MA: MIT Press, 1990), 64.

38. Williams, "Color Scheme," 278.

39. Greenhalgh, *Ephemeral Vistas*, 129.

40. John Daniel Barry, *The City of Domes: A Walk with an Architect about the Courts and Palaces of the Panama-Pacific International Exposition, with a Discussion of Its Architecture, Its Sculpture, Its Mural Decorations, Its Coloring and Its Lighting, Preceded by a History of Its Growth* (San Francisco: John J. Newbegin, 1915), 61.

41. Bernard R. Maybeck, *Palace of Fine Arts and Lagoon: Panama-Pacific International Exposition, 1915* (San Francisco: Paul Elder, 1915), 9. A model for the palace had been Arnold Böcklin's painting *Die Toteninsel* (The Island of Death).

42. *History of the Panama-Pacific International Exposition*, 39.

43. See ibid., 475–476, and Todd, *Story of the Exposition*, 3:369.

44. See Rose V. S. Berry, *The Dream City: Its Art in Story and Symbolism* (San Francisco: Walter N. Brunt, 1915), 24.

45. Juliet James, *Palaces and Courts of the Exposition: A Handbook of the Architecture Sculpture and Mural Paintings with Special Reference to the Symbolism* (San Francisco: California Book, 1915), 68.

46. Stella G. S. Perry, *The Sculpture and Murals of the Panama-Pacific International Exposition: The Official Handbook* (San Francisco: Wahlgreen, 1915), 3.

47. Haskin, *Panama Canal*, 374.

48. Cited in Perry, *Sculpture and Murals*, 3.

49. Katherine Delmar Burke, *Storied Walls of the Exposition* (n.p., 1915), 53. Another writer offered the explanation that the figures were of different sex "to indicate the dual nature of man urging him to greater triumph." Perry, *Sculpture and Murals*, 3–4.

50. Perry, *Sculpture and Murals*, 28.

51. Ibid.

52. Barry, *City of Domes*, 51.

53. In William H. Truettner, ed., *The West as America: Reinterpreting Images of the Frontier, 1820–1920* (Washington, DC: Smithsonian Press, 1991), 353, *The End of the Trail* is described as the best-known sculpture in the United States according to Fraser's obituary in the *New York Times*. The artist was only seventeen when he first envisioned the sculpture and spent a lifetime reworking it. At the

time of the exposition, he was almost forty. Benedict in *The Anthropology of World's Fairs*, viii, notes that the PPIE version had supposedly been destroyed but resurfaced in 1968 and then became a permanent exhibit at the National Cowboy Hall of Fame in Oklahoma City.

54. Sheldon Cheney, *Art-Lover's Guide to the Exposition: Explanations of the Architecture, Sculpture and Mural Paintings, with a Guide for Study in the Art Gallery* (Berkeley: At the Sign of the Berkeley Oak, 1915), 44.

55. Eugen Neuhaus, *The Art of the Exposition: Personal Impressions of the Architecture, Sculpture, Mural Decorations, Color Scheme and Other Aesthetic Aspects of the Panama-Pacific International Exposition*, 2nd ed. (San Francisco: Paul Elder, 1915), 32.

56. John Daniel Barry, *The Meaning of the Exposition* (San Francisco: H. S. Crocker, 1915), 19.

57. Perry, *Sculpture and Murals*, 44.

58. Macomber, *Jewel City*, 46.

59. Burke, *Storied Walls*, 68.

60. Perry, *Sculpture and Murals*, 86.

61. *Official Guide of the Panama-Pacific International Exposition San Francisco 1915*, June ed. (San Francisco: Wahlgreen, 1915), 34.

62. *Official Guide of the Panama-Pacific International Exposition San Francisco 1915*, Sept. ed. (San Francisco: Wahlgreen, 1915), 33.

63. Burke, *Storied Walls*, 66.

64. Ibid., 67.

65. *Looking Backward, Looking Forward*, n.p.

66. Neuhaus, *Art of the Exposition*, 73. His application of the term "rebirth" is similar to Roosevelt's concept of moral transformation (here: through art).

67. Macomber, *Jewel City*, 84.

68. Lears, *No Place of Grace*, 33.

69. See also Elizabeth N. Armstrong, "Hercules and the Muses: Public Art and the Fair," in Benedict, *Anthropology of World's Fairs*, 117.

70. Barry, *Meaning of the Exposition*, 12.

71. Stella G. S. Perry, "Shall I Go to the Panama Fair?" *Ladies' Home Journal* 32 (June 1915): 17. Women also contributed to the fair through the Women's Board, whose work was reviewed in Anna Pratt Simpson, *Problems Women Solved: Being the Story of the Women's Board of the Panama-Pacific International Exposition, What Vision, Enthusiasm, Work and Co-operation Accomplished* (San Francisco: Women's Board, 1915). Phoebe Apperson Hearst, wife of the newspaper magnate, served as the board's honorary president.

72. William MacDonald, "The California Expositions," *Nation*, Oct. 21, 1915, 491.

73. Todd, *Story of the Exposition*, 4:79.

74. Ibid., 4:385, 38.

75. See List of Moving Pictures Shown in the Various Exhibit Palaces, folder Motion Pictures, Panama-Pacific International Exposition Vertical File, San Francisco History Center, San Francisco Public Library, 3.

76. See Rydell, *All the World's a Fair,* 231.

77. Franklin K. Lane, "A City of Realized Dreams," *National Geographic Magazine* 27 (Feb. 1915): 170.

78. Todd, *Story of the Exposition,* 3:242.

79. Ibid., 2:376.

80. Ibid., 3:378.

81. Ibid., 3:376.

82. *History of the Panama-Pacific International Exposition,* 316.

83. Todd, *Story of the Exposition,* 4:40, 38.

84. Ibid., 4:38.

85. Ibid., 2:375.

86. See ibid., 5:41.

87. Ibid., 3:94.

88. Ibid.

89. See *Canal Record,* Aug. 5, 1914, 504, and Todd, *Story of the Exposition,* 1:170.

90. Neuhaus, *Art of the Exposition,* 73.

91. Edith K. Stellman, "With the Crowd at the Panama-Pacific Exposition," *Overland Monthly* 65 (June 1915): 519.

92. See Woody Register, "Everyday Peter Pans: Work, Manhood, and Consumption in Urban America, 1900–1930," *Man and Masculinities* 2 (Oct. 1999): 207–208. Thompson had also designed a ninety-foot caricature of a suffragette pejoratively nicknamed "Panama Pankaline Imogene Equalrights" but more commonly known as "Little Eva." See Benedict, Dobkin, and Armstrong, *Catalogue of Posters,* exhibit F2. Not far away, in the Palace of Education, the Congressional Union for Woman Suffrage presented the first suffragist exhibit at a world's fair. In California, the right to vote had been granted to women in 1911, and the suffragists were campaigning for a constitutional amendment.

93. Todd, *Story of the Exposition,* 2:374.

94. *History of the Panama-Pacific International Exposition,* 459.

95. See Todd, *Story of the Exposition,* 4:102.

96. *The Panama Canal at San Francisco* (San Francisco: Panama Canal Exhibition, 1915), 4. This booklet contains the official description of the concession.

97. See *The Blue Book, a Comprehensive Official Souvenir View Book of the Panama-Pacific International Exposition at San Francisco 1915* (San Francisco: Robert A. Reid, 1915), 309.

98. *The Red Book of Views of the Panama-Pacific International Exposition San Francisco 1915* (San Francisco: Panama-Pacific International Exposition Company, Robert A. Reid, Official Publisher of View Books, 1915), n.p.

99. *Panama Canal at San Francisco,* 5–6.

100. Ibid., 3. In the summer of 1914, Secretary Bishop was asked by Goethals to sail from Panama to the United States "for service in connection with the Panama Canal exhibit at the Panama-Pacific Exposition." It remains unclear to which exhibit Goethals was referring. It was Bishop's last assignment in almost nine years as secretary of the Canal Commission. Afterward his resignation went into effect, and Bishop resumed his literary career in New York. See *Canal Record,* July 1, 1914, 446.

101. *History of the Panama-Pacific International Exposition,* 460–461. It is unclear whether this number referred to revenues or profits. If we assume that about nine million visitors went to the fair (out of nineteen million total admissions, which included multiple entries), then—this is a very crude calculation— between 8 and 19 percent of them saw the exhibit (for the latter figure, the construction costs of $500,000 were added to the assumed profit to calculate the box-office receipts).

102. See Ewald and Clute, *San Francisco,* 109.

103. *Panama Canal at San Francisco,* 8.

104. Ibid., 7. The moving chairs (also with sound) would resurface at the General Motors Futurama exhibit at the World of Tomorrow fair in New York in 1939. See Nye, *American Technological Sublime,* 218.

105. *Panama Canal at San Francisco,* 7, 11; and Todd, *Story of the Exposition,* 2:150.

106. *Panama Canal at San Francisco,* 9.

107. *History of the Panama-Pacific International Exposition,* 460.

108. Stellman, "With the Crowd," 513.

109. *Panama Canal at San Francisco,* 4.

110. See Todd, *Story of the Exposition,* 2:150.

111. B. Brown, "Science Fiction," 143, links the "aerialized, globalizing point of view" to new ways of seeing, such as aviation and cinema. For another discussion of the Panama Canal exhibit, see Anne Maxwell, *Colonial Photography and Exhibitions: Representations of the "Native" and the Making of European Identities* (London: Leicester University Press, 1999), 92–93.

112. Bently Palmer, "World's Advance Shown in Exhibits," *Overland Monthly* 66 (Aug. 1915): 110.

113. B. Brown, "Science Fiction," 142–143.

114. McClintock, *Imperial Leather,* 58. On Foucault's notion of the panopticon, see also p. 230, n. 9.

115. B. Brown, "Science Fiction," 146.

116. Lane, "City of Realized Dreams," 171.

117. Charles C. Moore, "'San Francisco Knows How!': An Answer to the World's Question: 'Can This Exposition Be Different?'" *Sunset Magazine* 28 (Jan. 1912): 6.

118. B. Brown. "Science Fiction," 150.

119. See Maxwell, *Colonial Photography and Exhibitions*, 90.

120. *The Legacy of the Exposition: Interpretation of the Intellectual and Moral Heritage Left to Mankind by the World Celebration at San Francisco in 1915* (San Francisco: n.p., 1916), 62.

121. *History of the Panama-Pacific International Exposition*, 6.

122. *Why Americans Should See the Exposition, from Advance Sheets of Current Opinion for May* (n.p., 1915).

123. To prevent such damage, the leaflet *War — It Will Not Harm the Panama-Pacific International Exposition* (folder War, Panama-Pacific International Exposition Vertical File, San Francisco History Center, San Francisco Public Library) was printed shortly before the opening.

124. See Maxwell, *Colonial Photography and Exhibitions*, 93.

125. "The European War and the Panama-Pacific Exposition — A Monumental Contrast," *Current Opinion* 58 (May 1915): 315.

126. Todd, *Story of the Exposition*, 2:133.

127. On the 1939–1940 fair in San Francisco, see Lisa Rubens, "Representing the Nation: The Golden Gate International Exposition," in Rydell and Gwinn, *Fair Representations*, 121–139, and Robert W. Rydell, "The 1939 San Francisco Golden Gate International Exposition and the Empire of the West," in *The American West as Seen by Europeans and Americans*, ed. Rob Kroes (Amsterdam: Free University Press, 1989), 342–359.

Index

www.ingramcontent.com/pod-product-compliance
Lightning Source LLC
Chambersburg PA
CBHW060418100426
42812CB00030B/3226/J